HISTORY, PHILOSOPHY AND
SOCIOLOGY OF SCIENCE

Classics, Staples and Precursors

HISTORY, PHILOSOPHY AND SOCIOLOGY OF SCIENCE

Classics, Staples and Precursors

Selected By

YEHUDA ELKANA
ROBERT K. MERTON
ARNOLD THACKRAY
HARRIET ZUCKERMAN

THE FREEDOM
OF
SCIENCE

John R. Baker

ARNO PRESS
A New York Times Company

New York — 1975

Reprint Edition 1975 by Arno Press Inc.

Science and the Planned State was reprinted
from a copy in the University of Illinois Library
The Scientific Life was reprinted from a copy
in the St. Louis Public Library

HISTORY, PHILOSOPHY AND SOCIOLOGY OF SCIENCE:
Classics, Staples and Precursors
ISBN for complete set: 0-405-06575-2
See last pages of this volume for titles.

Manufactured in the United States of America

Library of Congress Cataloging in Publication Data
Baker, John Randal, 1900-
    The freedom of science.

    (History, philosophy, and sociology of science)
    Reprint of the 1943 ed. of The scientific life, and of
the 1945 ed. of Science and the planned state, both
published by Macmillan, New York.
    Includes bibliographies.
    1. Science and state--Addresses, essays, lectures.
2. Science--Social aspects--Addresses, essays, lectures
I. Baker, John Randal, 1900-   The scientific life.
1975. II. Baker, John Randal, 1900-   Science and
the planned state. 1975. III. Title. IV. Series.
Q125.B27   301.24'3   74-25150
ISBN 0-405-06636-8

CONTENTS

The
Scientific
Life

The
Scientific
Life

by

JOHN R.
BAKER

M.A., D.PHIL.,
D.SC.

Lecturer in Zoology in the
University of Oxford

New York
The Macmillan Company
1943

Preface

THIS book has been written for those people, scientists and non-scientists alike, who have the welfare of science at heart. In the period of reconstruction after the war the future of science will be decided. Science, long thought impregnable, is now attacked. The attack, subtle and indirect, comes from within. The spirit of free enquiry is threatened today by those who would plan other people's researches and confine investigation to crudely practical ends. If the threat materializes, the spirit of free enquiry will be crushed. That spirit has given mankind its heritage of modern scientific knowledge, and if it be allowed to live it will give new knowledge in ever-increasing measure.

Three wishes accompany this book. I hope that it will help to show research workers and teachers and students of science, and technologists too, that science is threatened. I hope that by its very shortcomings it will stir profounder minds to write on the same subject and thus bring substantial help to the cause of freedom in science.* My third hope is that some parent or science teacher may think fit to put this book into the hands of a thoughtful boy or girl aged fifteen or over, and that as a result that boy or girl will decide to live the scientific life. Before the last wish can be realized, freedom in science must be secured.

I am indebted to Professor A. G. Tansley, F.R.S., and Professor E. S. Goodrich, F.R.S., for kindly reading the manuscript and making many suggestions, most of which I have

* Professor M. Polanyi has now written a profound essay on "The growth of thought in society" (*Economica*, Nov., 1941).

adopted. Professor M. Polanyi and Mr. R. Brown have read parts of the manuscript and given valuable advice. Though not herself a scientist, my wife has made suggestions which have been of real value in most of the chapters. The librarians of the Radcliffe Library have shown never-failing helpfulness.

JOHN R. BAKER

DEPARTMENT OF ZOOLOGY AND COMPARATIVE ANATOMY
UNIVERSITY MUSEUM, OXFORD

Contents

Chapter 1

The Act of Discovery

"Lernen wir träumen, meine Herren, dann finden wir vielleicht die Wahrheit."   KEKULÉ[120]

THE word discovery is used in science with two meanings, though they are not always very distinct. On the one hand a new phenomenon or object may be disclosed, such as the penetrating powers of X-rays or a living specimen of a fish belonging to a group previously thought to have been extinct for millions of years: on the other a new principle or theory may be formulated, as when structural formulae were suggested in organic chemistry or the theory of natural selection put forward as a partial cause of evolution. Discoveries may usually thus be classified as factual or theoretical. In most sciences the factual discoveries are commonly made in the laboratory, but they originate in ways which would be surprising in their diversity to people who are not themselves concerned with scientific research. Ideas for experiments occur unexpectedly to the scientist anywhere and at any time of day or night, often when he is thinking of something else. Chance observations which he makes in his laboratory may suggest quite new lines of investigation. Important factual discoveries probably arise more commonly from such sources as these, than from the rigid following-up of a comprehensive plan of campaign deliberately thought out by the scientist in advance. Theoretical discoveries are seldom produced to order while the sanguine scientist sits conveniently confronted with a blank sheet of paper: they generally originate, like ideas for new lines of research, when he is far from laboratory or study.

Whenever one is able to look into the mind'of a discoverer at the moment of discovery, one sees that the finding out of unknown facts and the origin of great generalizations are not crudely mechanical processes, to be achieved by the efficient sorting of the cards in a card-index or the punctual study of pre-publication abstracts of other people's work. Ideas come in the most unpredictable way, and seldom when they are being sought. One distinguished scientist has told me that his new ideas often come during railway journeys, and several have mentioned the capacity of a hot bath to generate plans for new investigations or for theories to account for what is already known. In the case of Archimedes the startlingly sudden origin of his important hydrostatical concept following the immersion of his own body in his bath has perhaps misled people into assuming that he was thinking of himself as the body submerged, whose weight would lose by submersion as much as was equal to the weight of the bath-water displaced. It seems at least as likely that the bath was acting simply as a generator of ideas.

It need scarcely be said that whatever originates in the train or 'bus or bath or bed or whatever incalculable place it may be, must subsequently be subjected to the most searching analysis in the cold light of the laboratory; but the idea must come first. Let us watch a couple of ideas of the first magnitude coming into the mind of a great German chemist.

So long as people are still interested in the history of science the name of August Kekulé will be honoured, above all for his introduction of structural formulae in organic chemistry and his theory of the constitution of the benzene ring. He was not actually the originator of the concept of valency, but he devised the method of writing down the formulae of organic compounds which has been universally accepted ever since, on account of its demonstrating so clearly the way in which atoms and groups of atoms are linked together. The mere knowledge that so many carbon atoms, so many hydrogen atoms, etc., enter into a molecule of a substance does not mean very much; but if anyone with chemical knowledge can see, as he can from a structural formula, how they are linked, he can instantly infer many of the properties of the substance.

Kekulé himself has told us how the inspirations came which gave such a great impetus to chemical research. He announced it in his speech in the Berlin Rathaus on 11th March 1890, when the German Chemical Society met there to do him honour on the occasion of the twenty-fifth anniversary of the publication of his theory of the chemical structure of benzene[120]. After the delivery of speeches by notable German chemists and the reading of letters and telegrams in his honour from representatives of chemical societies in various parts of the world, Kekulé made a simple, modest speech in which he told how the two greatest discoveries of his life came to him.

He told how he lived at one time near Clapham Common, and how he often went to Islington to spend the evening with his friend Hugo Müller. One night the two friends had spent the evening together in this way, talking of many things, but mostly of their beloved chemistry. It was summer, and Kekulé went home on the outside of an omnibus. "I sank", he said, "into a reverie. The atoms flitted about before my eyes. I had always seen them in movement, these little beings, but I had never succeeded in interpreting the manner of their movement. That day I saw how two small ones often joined into a little pair; how a larger took hold of two smaller, and a still larger clasped three or even four of the small ones, and how all span round in a whirling round-dance. I saw how the larger ones formed a row and only at the end of the chain smaller ones trailed along.

"The cry of the conductor, 'Clapham Road', woke me up from my reverie, but I occupied part of the night in putting at least sketches of these dream-products on paper. Thus originated the structure-theory."

It was particularly the formulation of the theory of the benzene-ring that the assembly had met to commemorate, and Kekulé told how this discovery also came to him, a discovery on which one-half of organic chemistry is based and which has proved more fertile than any other discovery in making possible verifiable and true predictions about the properties of substances. Those who have not studied organic chemistry will scarcely be able to comprehend how Kekulé's theory of the structure of benzene revolutionized the subject. In its essence it was a very simple theory: the six carbon atoms of the molecule are arranged in a ring, each linked to the next. The idea of this ring-structure came to Kekulé in another of his reveries.

During his residence at Ghent, he was sitting writing his chemical text-book. "But it did not go well; my spirit was with other things. I turned the chair to the fireplace and sank into a half-sleep. Again the atoms flitted before my eyes." His imaginative eye, sharpened by repeated visions of a similar kind, could by this time distinguish large structures of complicated construction. He had seen rows of atoms linked together, but never yet rings: nor had anyone else. This is how the idea came to him: "Long rows, variously, more closely, united; all in movement, wriggling and turning like snakes. And see, what was that? One of the snakes seized its own tail and the image whirled scornfully before my eyes. As though from a flash of lightning I awoke; this time again I occupied the rest of the night in working out the consequences of the hypothesis."

Kekulé did not tell of these incidents simply for the sake of something to say; he was deliberately advising on how to make discoveries in science. "Let us learn to dream, gentlemen, then perhaps we shall find the truth"; but almost immediately he gave a warning against the publication of the dreams until they had been put to the test of the intelligence when awake.

As the embryologist, Dalcq[28], has insisted, it is wrong to consider sciences as constructions methodically erected according to a thought-out design. "Like all human achievements, they are a fruit of life, and of progress in those directions which happen, at a certain moment, to be favourable." One might add that those who contribute to science are drawn to their work not by any bureaucratic scheming, but by an imperious internal drive: if sufficient drive is there, it will overcome almost all obstacles. Every time a motorist speaks of an amp,* a wireless enthusiast of a microfarad, or a sailor of degaussing, a famous scientist is being commemorated who rose through sheer ability without the aid of wealth at a time when it was much harder to rise than it is today. To anyone with the intense desire to find out that characterizes the research-worker, any other life is tame, whatever attractions it may hold for a different type of person. A recent correspondent in "Nature"[83] has said that young scientists have sometimes concealed their degrees in science in order to get well-paid administrative jobs. He calls this "waste of material", but it may be suggested that this is quite the wrong comment. The men had had a training in science, but it is clear that they were not scientists; and the real waste would have occurred if some one had not realized their deficiencies and had given them posts

* Ampère's father was executed at the time of the French revolution. Still a youth, Ampère supported himself for a time by giving private lessons.

in research laboratories. One discovers or invents because one has an internal urge to do so, unless the environmental conditions make it impossible. Every help should be given to people in all walks of life who have the genuine urge. The help should be encouragement, facilities and a reasonable livelihood. The man who wants the higher pay of an administrative job should be dissuaded from occupying a potential research worker's seat in the laboratory of a scientific institution.

In an amusing advertisement in the "Daily Telegraph" a firm of tool-manufacturers[38] announce that they invent tools because they like inventing tools. This gives the impression of being one of the most truthful advertisements one has ever read: the inventor has an urge which is comparable to that of the discoverer. It is only genuine interest in a subject, and not the hope of reward, that can make a research worker accept willingly the continual rebuffs which he receives in his work. Continually things go wrong: repeatedly he discovers that weeks or months of hard work have been thrown away on a line that will lead nowhere, or some one else discovers the same thing without his knowing it, and it looks as though all the time he has spent on the subject had been wasted. The continual disappointments would cure all but the genuine natural investigator in a very short time of any desire to devote his life to research: he would be unlikely to get so far as experiencing the joy of even a very small discovery. The man who works for reward or fame would never be able to force himself to do all the "useless" reading in subjects not immediately connected with his research that is so often a genuine pleasure to the real scientist. He reads omnivorously when he is not using his hands in his laboratory, and he reads because he is interested. Much that he reads he will never use again: just every now and then a remembrance of something he read long ago throws light on his immediate problem: he has a new insight into his work. His wider reading is utterly different from that of the writer of a text-book: he need reject nothing because it seems irrelevant, and his wider reading—I am not referring here to his concentrated study of the literature obviously connected with his research—would seem sketchy to anyone who examined him on it. Meanwhile ideas are accumulating, and most of them will lead to nothing; a few, when he is lucky, to discovery.

Nothing but an intense interest in the subject as a whole —in things—could make a man work so long with so many rebuffs and read so much with so little hope of profiting directly. The youthful aspirant to fame, who reads of the honours showered on Pasteur, likes to think of himself as a great medical research worker. He forgets that Pasteur was no aspirant to fame, nor even interested at first in medicine. He was, like all great scientists, intensely absorbed in the study of things. It is useless to desire to become a research worker because one desires fame, and equally useless to do so because one thinks that one can serve the community by scientific research. The man who has the urge to find out serves the community indeed, culturally or materially or both, if given the opportunity; but he does so because he believes in the value of finding out the truth about things. A man with no special talents for music might as well say that he was going to compose a symphony because he owed it to the community, as one without the gifts of the natural investigator undertake scientific research for the same reason. No scientist can tell what he is going to discover: a man might seek something comparable in its beneficence to trichlormethane and end up with a substance so potentially horrible as $\beta\beta$-dichlordiethylsulphide to his credit.

The joy of discovery is a very real incentive to research, despite the rareness of its realization. It is an error to suppose that the scientist is unemotional, or could succeed if he were. The error has arisen through a misconception. The absolute necessity that a scientist's findings shall not be changed from objective truth in response to emotional urges of any kind does not result in his becoming a particularly unemotional person: whether a discoverer or anyone else is pleased with a discovery has no effect on its validity. "I have been working like a madman at Drosera", wrote Darwin to Sir J. D. Hooker in reference to his study of insectivorous plants, and a few days later, to the geologist, Lyell, "at the present moment I care more about Drosera than the origin of all the species in the world. . . . I am frightened and astounded at my results"[36]. Kropotkin once wrote[84], "There are not many joys in human life equal to the joy of the sudden birth of a generalization. . . . He who has once in his life experienced this joy of scientific creation will never forget it." Kropotkin did not exaggerate, and what he wrote is as applicable to factual as to theoretical discoveries. Let us witness an authentic case of joy in factual discovery.

"All is discovered!" The first shock is such that the young scientist cannot put his eye back to the instrument, but only emits this scarcely articulate cry. He rushes from his laboratory into the corridor. Meeting there a laboratory assistant in physics, he insists on embracing him. The assistant is dragged off to hear an exuberant account of what has transpired[136].

It was Pasteur who spoke. At the age of 25, before he turned to medicine, he made a discovery which, with those of van't Hoff, laid the foundations of the branch of chemistry which is concerned with the actual shapes of the ultimate particles of chemical compounds in the three planes of space.

It was already known that solutions of the salts of tartaric acid (from grape-juice) have a special effect on light. The light of the moon, or any other light in which the waves vibrate in one plane instead of (as in sunlight) in all planes at random, is affected by a solution of one of these salts. The light emerges from the solution with the waves still all vibrating in one plane, but the plane is a different one: the plane of polarization has been rotated to the right. Pasteur's discovery was this[72]. He found that when he evaporated a solution of a salt of an acid very closely allied to tartaric acid, two sorts of crystals were produced, differing only in the positions of certain of their surfaces, so that one kind might be called right-handed and the other left. One kind of crystal was a familiar one, the other quite new. Pasteur was able, with the aid of the microscope, to separate the right-handed from the left-handed crystals, though the differences were small and the possibility of the existence of such differences not previously envisaged. He dissolved each kind of crystal separately and passed a beam of polarized light through each solution.

What he saw was the cause of Pasteur's excitement. One solution changed the plane of vibration in the usual way, to the right: the other, made from the new kind of crystals, changed it to the opposite direction, to the left. Thus the symmetry of the crystals was correlated with the optical behaviour of their solutions, and an insight was given into the structure in the three planes of space of the ultimate particles of which chemical compounds are formed.

The excitement and joy of making a discovery is known to every research worker, though in Britain the expression of the emotion is more reserved, so that when our colleague discovers something and we play the part of the laboratory assistant in physics in the corridor, we avoid the necessity of being kissed. The emotion is there whether the discovery is big or small, whatever the nationality of the discoverer may be. Let us watch a very great British scientist making a very small discovery.

Alfred Russel Wallace, co-founder with Charles Darwin
of the theory of evolution by natural selection, catches a
new species of butterfly in the Moluccas. It is a particularly
brilliant species of the genus Ornithoptera. "None but a
naturalist", wrote Wallace[140], "can understand the intense
excitement I experienced when I at length captured it. On
taking it out of my net and opening the glorious wings,
my heart began to beat violently, the blood rushed to my
head, and I felt much more like fainting than I have done
when in apprehension of immediate death. I had a headache
the rest of the day, so great was the excitement produced
by what will appear to most people a very inadequate cause."

The joy of a discovery, big or small, is shared by a scientist's colleagues. The enthusiasm is infectious. Pasteur's
crystallographic discovery illustrates the fact very pleasantly.
He asked the old French chemist, Biot, for permission to
demonstrate his results to him. Biot agreed, insisting that
his own materials should be used and the test made in his
own laboratory. The solution was made and allowed to
evaporate for a couple of days. When sufficient crystals had
appeared, Biot called Pasteur once more to his laboratory
and in the former's presence the young scientist separated
the right-handed crystals one by one from the left-handed.
Biot dissolved them and decided first to examine the solution
which should, according to Pasteur, rotate polarized light
to the left. This was the solution of the new kind of crystal,
with hitherto unexpected optical properties. Biot called
Pasteur once more to his laboratory and proceeded to observe
the effect on polarized light. He knew at once that a great
discovery had been made. He seized Pasteur's arm and
exclaimed, "My dear child, I have loved science so much
in my life that that makes my heart throb."[136]

It was the sudden birth of a generalization—a theoretical
discovery—that Kropotkin held to be so particularly joyful.

Pasteur's discovery was factual, but not less satisfying for that. Kekulé's reveries resulted in theoretical discoveries. We may attend with advantage at the birth of another theoretical discovery, this time in biology, as important for science and as satisfying to its maker as Kekulé's.

Wallace himself regarded his work in describing and working out the distribution of the insects which he collected in the East Indies as incompletely satisfying or "comparatively profitless", as he expressed it[92]. We should realize that he was comparing the "profit" with that of his great work on evolution by natural selection. The circumstances under which the great generalization was born in Wallace's mind are known. To Charles Darwin—originator of the phrase "natural selection"—the same great generalization came very slowly: upon Wallace, to use his own expression, it "flashed". One is reminded of Kekulé's comparison with a flash of lightning, when he conceived the structure of the benzene ring. Wallace and Darwin discovered independently the theory of a cause of evolution which is still, in an improved form, the most widely held today and has been a great stimulus to research.

Wallace was in his bungalow in the little island of Ternate in the Moluccas, his headquarters for the study of the fauna of the East Indies. Here, amid all the treasures of natural history that he had collected, the great naturalist was taking an enforced holiday as a result of repeated attacks of malarial fever. It was during an actual attack of fever that the idea came to him. His mind was reflecting on Malthus's "Principles of Population", and he brought his remembrance of this book, which he had read twelve years before, into connexion with the vast stores of knowledge that he had gained of the lives of wild animals in their native haunts in the East Indies. It was a supreme example of the value of wide reading to research workers. The principle of the survival of the fittest "suddenly flashed" upon him. "Then at once", he wrote, "I seemed to see the whole effect of this", and he waited impatiently for his fit of fever to leave him, so that he could write down a sketch of his theory. That same evening he did so, and during the next two evenings he wrote out a fuller account to send to Darwin.

The scientist would be a self-centred person if he cared nothing for the good opinion of his colleagues. It is not fame that he wants, for that would mean that he respected a value set upon his work by those who could not be in a position to assess it truly. The good opinion of colleagues is desired by nearly every research worker, and when it is gained—and it can only be gained by the revelation of demonstrable truth—there is a satisfaction as great as and perhaps more lasting than the initial joy in discovery, a satisfaction all the more real because others can take genuine unselfish pleasure, like Biot, in the discovery. Ray Lankester expressed these feelings very directly when he gave an account of the first known fresh-water jelly-fish.

One Thursday in the summer of 1880 the Secretary of the Botanical Society of London noticed some strange organisms floating in a warm-water tank, in the house in Regent's Park devoted to the cultivation of the magnificent water-lily, Victoria regia. By the following Monday he had placed a number of specimens at Lankester's disposal. The organisms were jelly-fish. Never previously had a jelly-fish (Medusa) been known to inhabit fresh water: the kinds of animals popularly so called were thought by the best scientific opinion of the time to be exclusively marine. Lankester wasted no time; nor did Allman[4], another distinguished zoologist, to whom the Secretary also gave some specimens. On the Thursday morning, just one week from the time the organism was first seen, a description of it by Lankester was already in print[85], and on the same afternoon Allman was describing it at the Linnean Society. (By the inflexible law of priority, Lankester's name of Craspedacusta was accepted for the animal in preference to Allman's Limnocodium, though Lankester, with a somewhat Irish gesture, tried simultaneously to change his name into a philologically more exact form and to withdraw it altogether out of respect for Allman.) A week later quite a long paper on the physiology of the new animal was already published by the celebrated biologist whose memory is kept fresh in Oxford by the Romanes lectures, still delivered annually on scientific or literary topics[116]. On the same day Lankester wrote[86]: "I confess to having worked at that Medusa day and night when I first obtained it, with the object of having the pleasure and honour of being the first to expound its structure to my brother naturalists."

Lankester referred to the pleasure and honour of his work; Darwin in his autobiography put love of science first in the list of the mental qualities which gave him success in research[36]. It may be left to a philosopher to describe more particularly the value of science to the scientist. The pleasure of science, as Alexander[3] says, "is sometimes felt passionately, most often it is a calm delight in contemplating the harmonies of knowledge. It may be attended by subsidiary excitements in the work of investigation, some of which are pleasurable in so far as the labour tends to success; others involve pain or suspense. The release from such tension, when disappointment and frustration are replaced by discovery, adds a glow to the exercise of the search. . . . Truth is the satisfaction of disinterested curiosity."

Chapter 2
Scientists as People

ONE day I was sitting at a window with a friend, when an eminent scientist pedalled sedately by on his bicycle. My friend was unacquainted with the personalities of science. I remarked that we had just seen a particularly distinguished Fellow of the Royal Society. "What?" he exclaimed, "a Fellow of the Royal Society? He looks more like the man who comes to mend one's umbrella." Never having been visited by anyone anxious to perform this particular service, I could not form an exact idea of how the scientist appeared to my friend; but it was clear that the impression was unfavourable. The public has certain ideas of what a scientist should look like and how he should behave. As Waddington[139] has recently remarked, the popular idea of a scientist is of someone removed from ordinary human realities. On reflection, it seems strange that there should be such a thing as a popular idea on the subject at all, for the great majority of human beings have not only never themselves made the smallest contribution to knowledge in the scientific sense, but have never known personally anyone who has. Nevertheless, there is some truth in the idea described as popular.

In ordinary life loyalty and steadfastness in opinion are admired. A scientific friend of mine was described by another scientist as "loyal to his professor". The accusation of being loyal to his professor's scientific views was intended and received as a deliberate insult, for scientists acknowledge no loyalty to the opinions of persons, but only to the truth as it appears to themselves. They give honour to one another for change of opinion. In 1855 T. H. Huxley strongly attacked what he called the "fallacious doctrine"[78] of evolution. Five years later his opinion had completely changed, and he delivered his famous and shattering attack on the Bishop of Oxford, who sought to ridicule Darwin's views. The episode occurred in a room in the University Museum at Oxford, and there is now a plaque outside the door to commemorate it. The Bishop had taunted Huxley on his belief in his own (Huxley's) descent from an ape. Huxley's reply is well known. "I asserted, and I repeat," he said, "that a man has no reason to be ashamed of having an ape for his grandfather. If there were an ancestor whom I should feel shame in recalling, it would be a *man*, a man of restless and versatile intellect, who not content with an equivocal success in his own sphere of activity, plunges into scientific questions with which he has no real acquaintance, only to obscure them by an aimless rhetoric, and distract the attention of his hearers from the real point at issue by eloquent digressions, and skilled appeal to religious prejudice"[36]. The doctrine of evolution no longer appeared fallacious to Huxley, and he was respected for changing his views.

The eminent geneticist, Bateson, opposed the chromosome theory of heredity almost throughout his working life. Not long before he died he visited America, saw for himself the chromosome preparations that had convinced others long before, and changed his mind. "I am heartily glad I came", he wrote from America[14]. "I was drifting into an untenable position which would soon have become ridiculous." Bateson was respected for his change of opinion.

The popular idea that scientists are often absent-minded is not wrong: it only means that they are present-minded on something other than the trivialities of ordinary life. From a child onwards van't Hoff was fond of daydreaming[47]. When he was twenty-two he formulated a theory which became one of the foundations of stereochemistry. He could not have done that if his mind had been much concentrated on the little matters of his immediate surroundings. Kekulé's dreams have already been mentioned; the 'bus conductor probably found him absent-minded when the idea of structural formulae was forming in his brain. Bateson has related how "having supposed I had a ticket for "The Pillars of Society" I found myself at a musical extravaganza by some mistake". The very next night he went to hear "Siegfried", but "unfortunately I was half an hour late through mistaking the time". His pencil, notebook, knife, forceps, scissors, pipe and spectacles were perpetually mislaid. He would go to London in old "garden" flannels, darned across the knee, but wear a brand-new "town" suit in the garden and kneel on the gritty path to look at his plants[14]. This kind of behaviour is commoner among research workers than in most walks of life. We are told[80] that at an interview with a young scientist Einstein was dressed in a morning coat and "striped trousers with one important button missing".

If the scientist tends to be rather peculiar in an inoffensive way, then the nature of his work often gives him more than an ordinary share of certain virtues which are also common outside science. A sort of kindliness is engendered by continued observance of cause and effect, and savage hates are foreign to the scientific mind. A spread of scientific culture would lessen racial animosity and thoughtless dislikes of many kinds. It is strange to the mind of a scientist that anyone should want to persecute a person because he is a Jew, a communist or a kulak. The scientist's attitude is one of interest in the ethnological, psychological or sociological circumstances which produce those three categories of persons. (Incidentally, the debt of science to Jewish research workers is so great that the thought of the persecution of their race is particularly abhorrent to scientists.)

The great mountaineer, Whymper, was not a scientist. He devoted some pages of a perfectly serious book[144] to the cretins and sufferers from goitre in the valleys of the Alps. He was horrified by these people in a way which would be impossible to a scientist, who would regard the diseases impersonally and look for the cause. Whymper allowed his horror to grow to hatred, and wished to think that the sufferers or their parents were themselves to blame. He wanted those suffering from goitre to be conscribed into monstrous armies, to be commanded by idiot cretins. He wanted illegitimate cretins to be subject to "special disabilities", as though the frightful disability of cretinism were not a sufficient punishment for innocent persons. That people should think in this way appals the scientist. Of a great geneticist, for instance, it was written, "There was no place for punishment in his philosophy"[14]. Research has exhibited the folly of the mountaineer. Lack of iodine in the drinking water, not wickedness, is the cause of the diseases. One ounce of iodine in seven and a half million gallons produces a drinking-water which will ward off goitre[146], while feeding with thyroid gland or its extract has a markedly beneficial effect on cretinism.

Hand in hand with the lack of hate towards others there sometimes goes an almost childish unwillingness to believe in the guile or hatred of others. An incident in the life of Archimedes illustrates this kind of unworldliness. When Syracuse was besieged by the Romans in 212 B.C., Archimedes put his great inventive power at the disposal of the defenders. It was in vain, for the city was taken by surprise. Exactly what happened is uncertain[63], but it would appear that Archimedes was absorbed in the solution of a geometrical problem and did not notice that the Roman soldiers had arrived. When one of them approached to take him prisoner, he asked for time to solve his problem. The soldier answered by drawing his sword. A moment later the great mathematician and scientist was dead.

Modesty is another virtue particularly common in scientists. The work itself and the appreciation of colleagues are the only rewards desired. Collip, who played an important part in the research which gave the world insulin, wrote these words: "The part which I was able to contribute subsequently to the work of the team was only that which any well-trained biochemist could be expected to contribute, and was indeed very trivial by comparison with Banting's contribution"[22]. The modesty of these words is impressive, and a good example of a trait which is probably commoner in the scientific than in most professions. Another good example of modesty is furnished by Ernst Abbe, the great German computer of microscope lenses. When he introduced apochromatic lenses and thus revolutionized critical microscopy, his paper was so written as to "avoid even hinting at the almost gigantic work of calculation" which underlay their design. As one who knew him well remarks[115], "an attentive reader will see that Abbe was totally innocent of any tendency towards self-praise." Among scientists conceit and arrogance are rare.

There is a simplicity about most scientists, especially great ones, which is attractive. The discoverer does not wish to pose as an encyclopaedic slot-machine: indeed, those whose knowledge is greatest are seldom the most successful in research. I shall always remember the wise words which were spoken to me many years ago when I was about to address the Royal Society. The Secretary took me aside just as we were about to enter the lecture-room. "Now remember," he said, "you are going to speak to eminent scientists; so you must speak very simply." Had they devoted their lives to the cultivation of omniscience, they could not have become eminent in scientific research.

An incident in the life of Pasteur shows conclusively that great knowledge is not necessarily required for success in research. The great scientist was persuaded, despite his ignorance of the subject, to study the diseases which were becoming a menace to the French silkworm industry. On arrival at the region where the epidemics were at their worst, he met the French naturalist, Fabre, and asked him for information about the natural history of the silkworm moth. Fabre has recorded the conversation which ensued[46]. The great scientist was not aware that the cocoon contained a transformed larva called a chrysalis. Six years from then the diseases were conquered and the silk industry of France saved, thanks to Pasteur's investigations; and an important step forward had been taken towards the understanding of the relation between germs and disease in general. At the start he had not known a fact familiar to most children. His ignorance made a deep impression on Fabre, who knew and admired Pasteur's work in crystallography and bacteriology. Pasteur's "magnificent assurance impressed me", he wrote; "I was astounded; more, I was filled with wonder." Subsequently he deliberately made it a rule to adopt what he called the "method of ignorance" in his own entomological researches.

Great scientists are often simple in another way: they prefer simple apparatus. In these days there is sometimes a sort of competition between institutions in the size and cost of their more spectacular instruments. The cyclotron is a very large and very expensive apparatus used to accelerate electrified particles in experiments on the bombardment of atoms. Each institution anxiously announces the tonnage of its own instrument, like a shipping firm intent on impressing the public by the magnificence of its property. The greatest discoverers have acted differently. Von Baeyer, the brilliant German synthetic chemist known above all for his researches

on indigo, always worked with the simplest possible apparatus, mainly test-tubes, glass rods, small flasks and beakers.
He used no large piece of apparatus and no elaborate
mechanical device. Perkin, the great British chemist who
revealed the chemical constitution of camphor and of various
natural dyestuffs and alkaloids, also used the simplest equip-
ment[64]. Wollaston, discoverer of two elements and of the
dark lines in the solar spectrum and inventor of the reflecting
goniometer, carried simplicity still further than Baeyer and
Perkin. A foreigner once called and asked to see the great
scientist's laboratory. It was not necessary to move from
the room to do so. Wollaston instructed his butler to bring
his laboratory, and the order was obeyed—on a tea-tray[55].

Scientists are more comparable to musicians and artists
than to business men and politicians, and it frequently
happens that they are themselves musical or artistic. Galileo,
for instance, was a skilful amateur musician, and Perkin was
equal to a professional in this respect[64]. Anyone who has
witnessed a performance of "Prince Igor" and been thrilled
by the amazing vitality of the musical accompaniment to
the Polovtsy dances may find it difficult to reconcile what
he has heard with the sober fact that the composer, Borodin,
was assistant professor of chemistry at St. Petersburg
Academy of Medicine and later lecturer in the same subject
at the school of medicine for women which he himself had
helped to found. Rimsky-Korsakoff has told how Borodin
would suddenly leave a conversation on music in his house
to dash along to the adjoining laboratory, making the con-
necting corridor ring with some extraordinary passage of
ninths or seconds. Having assured himself that everything
was well in the chemical sphere, he would return to the
house for more talk and music[83a].

Scorn has recently been poured on the pursuit of pure
science as an "elegant pastime". Some of the greatest scien-

tists have openly practised elegant pastimes. Perkin was an expert on Malmaison carnations[64]. Many scientists adopt a science other than their own for spare-time interest. Armstrong, the chemist, was interested in the geological basis of scenery[8]: the discoverer of the penetrating powers of X-rays was a student of Alpine flora during his holidays[52]. Those who scorn "elegant pastimes" are apt to think scientists childish, because neither children's play nor the pursuit of knowledge has immediate application to material affairs. Scientists will not be horrified at being regarded as childish; they regard the "childlike simplicity and unworldliness"[51a] of Faraday as a virtue, not a vice. The enthusiasm of a child remains as the enthusiasm of an old scientist, while others flatter themselves on being grown-up. The interest of a child in his environment remains as a scientist's interest in nature: he does not undergo a metamorphosis into a different kind of creature, as though he were a chrysalis turning into a butterfly. The good scientist does not pretend that his interest in things is of a different kind from that which he experienced when a child. He has the satisfaction of knowing that it was the assiduous cultivation of such interests by generations of scientists that gave the world its cultural and material treasures of knowledge. The actual interests of childhood often persist in both science and technology. A boy called Sidney Camm was enthusiastically interested in model aeroplanes: as a man, the designer of the Hurricane fighter, he saved his country in the Battle of Britain[41]. Hogben has announced[71] that when he became a man, he put away childish things; but this is unusual in discoverers. An adult scientist has been known to spend a quarter of an hour playing with the "up" and "down" buttons of an electric lift, "thoroughly enjoying himself, laughing delightedly like a youngster at play"[145]. This will scandalize those who think that a scientist should be quite a different sort of human being from the child from whom he developed. It is perhaps because the scientist on the lift retains some spirit of childishness that he is such a very great man. It was Albert Einstein, not long before he was forced to leave Germany.

The scientist likes using his hands, as many children do; he could not succeed in his work if he did not. He does not claim that his childish enjoyment disappeared at maturity and was replaced by an entirely different phenomenon. He knows that his childish manipulations developed steadily into those which serve him so well in the laboratory. He is immune to the sneers of people who say that he has become a scientist because he wants a "legitimate excuse to engage in enjoyable tinkering"[26]. People sometimes jealously express their disapproval of scientists being allowed to do what they enjoy doing in their laboratories, while so many have uncongenial occupations. Those who profess these sentiments do not know what they are talking about. The majority of people, if they could glimpse the future, would view with nothing less than horror a condemnation to a life-time of scientific research. They would shrink from what they would regard as the unutterable boredom of it, and from the repeated disappointments which are a necessary concomitant. Ambitious people would do everything in their power to avoid a life in which success is generally recognized only by a small circle of fellow-workers.

The true scientist does not want public applause. He is a curious mixture of extrovert and introvert. In so far as he necessarily pours out his libido on objects external to himself, he is an extrovert; and certainly he does not turn his mind in upon his own soul in the manner of a character in a Russian novel. A certain withdrawal from the superficialities of everyday life is, however, almost necessary if the internal mental life is to be given a chance to produce results. He does not find satisfaction in the direction of others: he wants quietude to think. In these respects he is an introvert.

Some degree of withdrawal from social life is recorded again and again in the biographies of great scientists. Galen, one of the early exponents of true scientific method, expressed several times in his writings his scorn for people who spend their time going about saluting their friends[129]. Wollaston, again, to take a random example characteristic of many, is described as a silent, austere man, living only for his work, seldom taking part in social life and then only at the Royal Society or Royal Society Club[130]. Of Graham, the originator of the scientific study of colloids, it was written that he was "Too retired, too quiet. . . . Very intimate friends he had few"[119]. Darwin felt confused in a large gathering: he was oppressed by the numbers of people at Royal Society soirées[36]. Nine days before his marriage he was wondering, in a letter to his future wife, why he "should so entirely rest my notions of happiness on quietness and a good deal of solitude"[87]. He almost seemed to welcome ill-health, which "has saved me from the distractions of society and amusement". "We have given up all parties," he wrote, "for they agree with neither of us." Faraday "took little part in social movements, and went little into society"[51a]. Instances like these could be multiplied. Mendel was "extremely reserved" and had very few intimates[79a]. He lived in a monastery, grew peas in the strip of garden beneath its little clock-tower, addressed the Brünn Society for the Study of Natural Science, and made discoveries in genetics which dwarf everything else that has ever been done in that science before or since, a science which is not only of the utmost academic interest, but is already, despite its youth, playing a large part in practical affairs.

It has been written of Dirac, the eminent physicist whose book on quantum mechanics has been compared in importance with Newton's "Principia", that "His loneliness and shyness were famous. . . . Only a few men could penetrate his solitude"[80]. Similarly Einstein "remains lonely, loving solitude, isolation and conditions which secure undisturbed work". He suggested jobs as lighthouse-keepers for refugee scientists, so that they might have the isolation necessary for scientific work. In the present century, as in former times, some of the greatest scientific work continues to be done by those who are by nature solitary. "I am a horse for single harness", wrote Einstein, "not cut out for tandem or team-work"[80].

One of the most extreme instances of unsociability that is to be found in human history is provided by a scientist. This man was known to flee from a company of strangers uttering a queer cry like a frightened animal[81]. He took his solitary walks after sundown, so as not to be observed, and made unusual sounds while so employed. He was seriously described by an acquaintance as more silent than the Trappist monks. He had his library four miles from his home, so as to avoid unnecessary meetings with people coming to consult his books[7]. (Incidentally he allowed his friends free use of his library, but would not permit himself to take out a book without leaving a receipt[76a]). He once wrote that to increase his acquaintance was "the thing which I chiefly study to decline"[26]. With his unsociability went a childlike unworldliness. When a plea was made to him for the relief of one of his librarians during illness, he had no idea how much money he should give, and offered £10,000. When dying, as when living, he wished to be alone. His wish, expressed as a positive order to one who would stay with him, was granted[130].

The reader may say that this was not a scientist, but a madman. He happens to have been one of the greatest scientists the world has ever known. Henry Cavendish

(1731–1810) was the discoverer of specific heat, one of the earliest investigators of latent heat, the first serious student of the properties of hydrogen and carbon dioxide, discoverer of the chemical composition of water, discoverer of nitric acid, discoverer (before Coulomb) of the inverse-square law in static electricity, and the first to get an approximation to the density of the earth by experiment.

Our modern scientific moralists are fond of telling us of the wickedness of academic scientists, who are often rather retiring people and somewhat inclined to shut themselves up in their laboratories away from the outside world. These moralists speak of the "integrated" life, the full life, the life which they think everyone ought to lead, scientist and non-scientist alike. This attitude is part of their general desire for uniformity and hatred of individualism. It never occurs to those who profess the integrated life to reflect on the history of science and notice what a large share in scientific discovery has been played by people who could lay no claim to integration. Great discoverers rarely have the inclination, or indeed the ability, to practise the "integrated" life, excellent as this ideal may be for those who are not absorbed in scientific investigation.

There are those who think that quietude, seclusion and freedom may be allowed to geniuses, but not to lesser minds. It will be recollected that Pavlov, the celebrated Russian physiologist, was allowed to avoid planned science and to exist as a sort of museum-piece of freedom in a totalitarian state. Before it is decided to adopt such a practice in our own country, it would be well to consider the nature of scientific genius.

Only a superficial person or one suffering from feelings of inferiority will deny that persons of exceptional talent do exist. A thoughtless person, having once grasped the principle of the structure of the benzene ring, may deceive himself into imagining it so simple that anyone might have thought of it. This would reveal historical ignorance. Let anyone who is inclined to think such thoughts engage himself in the heavy task of forgetting everything he knows about chemistry except what was known when Kekulé put forward his theory. Let him study in detail the conflicting evidence available at that time as to the structure of the benzene molecule. Let him remember also that at the present moment there must be a multitude of concepts of equal importance and simplicity in science simply calling out for recognition. Most scientists cannot see them and prefer to plod along, checking here, reinvestigating there, following blindly the safe old tram-lines.

Geniuses, then, exist; but their recognition presents a difficulty: the less one knows about a subject, the easier it is to recognize the geniuses. With a little knowledge, one rattles off their names with confidence. In chemistry, for instance, omitting people alive today, one may unhesitatingly enumerate Priestley, Lavoisier, Dalton, Avogadro, Graham, Kekulé, Mendeléeff, van't Hoff, Arrhenius. If one knows something about a subject—let us say biology—the difficulty is more considerable. One begins with Darwin, certainly, and Mendel; but then trouble begins. The trouble is not caused by lack of great men, but by the profusion of them. One realizes that at certain times certain possibilities of discovery were "in the air"; that if one person had not made the discoveries, another, nearly as good, would probably have done so. Here is an obvious distinction from a kind of genius that occurs outside science. If Mozart had not composed that immortal work of genius, the overture to "Le Nozze di Figaro," no one else would ever have done so; but if Kekulé had not lived, structural formulae and the benzene ring would not have remained for ever hidden: someone else would eventually have dreamed the same dreams.

The genius in science is not a man apart. If intelligence could be measured statistically in such a way as to include the extremes, it is scarcely to be doubted that the results would be displayed by an ordinary variability polygon, with the great majority of people possessing ordinary ability of various degrees and fewer and fewer in each grade as the extremes of genius and idiocy were approached. Those who are less than geniuses are very important indeed in scientific research: they make an enormous contribution to discovery. If Bach, Mozart, Beethoven and Wagner had never lived, music would be immeasurably poorer: there have never been four chemists whose work was so essential for chemistry as theirs for music.

Those who are less than geniuses in science are comparable with geniuses in their mental make-up in matters not obviously connected with science. There is the same necessity for a reasonable degree of quietude and separation from the stress of affairs, so that the mind may be open to receive the original ideas which come so rarely and are the basis of discovery. In Great Britain science occupies only one or two persons in ten thousand. Many scientists are wholly engaged in industry and teaching. Only a few hundreds altogether in the whole country are research workers in pure science. Does the community want to dry up discovery at its source? If not, it can afford to grant some degree of quietude to its investigators.

It may seem paradoxical, but the research worker sometimes can and does obtain the necessary mental quietude while bombs are heard to fall outside; but he cannot obtain it if he throws himself heart and soul into the conflict of politics, where all his values, above all his regard for truth and open-mindedness, are out of place, where injustice is as a matter of course condoned as expediency, and where friendship for those with whom one disagrees may be misconstrued as disloyalty. All this makes an atmosphere which is uncongenial to the research scientist, and it would be beside the point to tell him that he "ought" to put up with what is uncongenial. The community should ask itself whether it wants the scientist to discover or not: if so, let it give him the tools, and he will do the job. His wants are small: he should have a reasonable livelihood and a reasonable share of quietude, and he will make cultural and material return by discovery and by stimulus to other discoverers; but one might as well take the spade from a gardener as a reasonable degree of seclusion from a research worker. If a young scientist's real interest is seen to lie in politics, one has but a single duty to him and to the community: he should be encouraged to enter politics. Science will survive only if it is served by those whose true interest is the search for truth.

Despite the obvious difference in outlook between the scientist and the politician, it is becoming fashionable to urge scientists to study politics and to express surprise that there is only one scientist in the House of Commons. Scientific knowledge is not so important for administrators as might be thought. Administrators and members of the government should have a scientific background, a knowledge of the value, spirit and method of science; but they need not be scientists. A non-scientist who has seen a man die of phosgene poisoning is as good a judge as the most expert chemist or pathologist of the desirability of the use of gas in warfare. Scientists who take to politics are apt to leave the scientific method behind them in their laboratories, and immediately begin to use phrases and slogans of a looseness and incomprehensibility which would shock them to the core in their proper subjects. Thus, they will write glibly of our "governing class", as though those two words conveyed any intelligible meaning in present-day Britain. Waddington leaves his laboratory and proceeds to tell us that living things grow, reproduce and evolve; that these are processes, not "static things"; that only Marxists have really taken notice of these facts; and that if anyone does take notice of them, he is an adherent to the Communist doctrine of dialectical materialism. The reader does not need to be told that this is nonsense, but he is asked to refer to p. 81 of Dr. Waddington's book to see that I have not misconstrued what he says[139].

In his history of technology, to which for some inexplicable reason the name of "The social relations of science" is attached, J. G. Crowther[26] pours scorn on people who, according to him, regarded certain lines of investigation as "respectable" and others as "socially disreputable"; but he does not realize that in putting all the emphasis on the applications of science he himself is trying to make a part of science—the most solid and fruitful part—"socially disreputable". He claims that "the greatest service that can be rendered to science in a period of crisis is to assist the struggle of the progressive class for power". On the contrary, the greatest service that can be rendered to science in a period of crisis or at any other time is to give opportunity to those talented people, of either sex, of every class, of all races, of any religion or none, who have that genuine urge to find out and generalize which is the basis of discovery.

Chapter 3
Individualism in Science

"The grand, leading principle, towards which every argument unfolded in these pages directly converges, is the absolute and essential importance of human development in its richest diversity." WILHELM VON HUMBOLDT[76]

IN the past, all reformers have been anxious to improve the lot of the less fortunate members of the community, and they have wished to do so by spreading freedom more widely. Now a new kind of reformer has sprung up, who wishes to improve the lot of the community by the opposite process of taking freedom away. In the past, the progressive has always worked for freedom; now, the reactionary agitator pretends to be a progressive. It is an astonishing fact that even science has provided its quota to the ranks of those who are bent on reducing freedom, and more astonishing still that even the reduction of freedom in science itself is contemplated.

J. G. Crowther[26] writes that a factor disposing scientists to acquiesce in dictatorship is their habit of accepting authority in their own work. This is a strange statement. Scarcely a single scientist in any part of the world will agree that scientists in general are so disposed, for science is precisely that subject in which authority counts for nothing. The motto of the Royal Society (*Nullius in verba*) states this explicitly, and every scientist knows that discoveries must be demonstrable, for no one will take them on anyone's authority.

As Bertrand Russell has said[118], men who like administration think that it is good for the populace to be treated
like a herd of sheep. Nowadays there are those who think
it good not only for the populace but also for scientific
research workers. This is the message of Bernal's "Social
function of science"[15]. Anyone who reads this book must
become aware that it preaches a doctrine of the reduction
of freedom for research workers. It is true that here and
there the book contains passages which pay lip-service to
the ideal of freedom; but these passages are contradictory
to the whole tenor of the book.

Bernal does not stand alone. Waddington[139], who tells
us three times in a small book on "The scientific attitude"
that totalitarianism is bound to come to Britain, is faintly in
favour of individualism on one page and faintly opposed to
it on another. He tells us on one page that communism is
"near" to science, and on another that for communism the
final test of value is not critical experiment but service to
the interests of a particular class of the community; loyalty
to that class is more important than truth. What "near"
means under these circumstances, it is impossible to guess.
It is only fair to say that Waddington admits the necessity
for every scientist to be free to disagree with the majority,
including his elders and betters. Scientific progress, as he
says, would come to a "fairly immediate" end if that were
not allowed.

J. G. Crowther[26] suffers from none of Waddington's nice
hesitations about ushering in the totalitarian régime for
scientists. He discusses whether freedom is better or worse
than an inquisition instituted to make sure that scientists
hold the proper political views. He finds nothing good or
bad about freedom or inquisition in themselves, but claims
that whichever of them supports a progressive class is good.
Scientists must therefore be subjected to a political inquisi-

tion if that serves a particular class of the community which is called "progressive". What godlike creature decides whether a given class is progressive, we are not told. The reception of this book shows that there has been an extraordinary change in people's attitude to liberty. In 1916 R. A. Gregory wrote a stimulating book[55] on the true spirit of science. "The men who have advanced the human race throughout the ages", he said, "are those who have stood for individuality as against the conclusions of the crowd." A quarter of a century later, now Sir Richard Gregory, F.R.S., President of the British Association for the Advancement of Science, he writes a favourable review[56] of Crowther's book.

As contempt for liberty grows, so everything that distinguishes one person from another becomes a subject for scorn. "Originality", as J. S. Mill wrote[98], "is the one thing which unoriginal minds cannot feel the use of." Not only can they not feel the use of it: they hate and fear it. Geniuses are "more individual than any other people—less capable, consequently, of fitting themselves, without hurtful compression, into any of the small number of moulds which society provides in order to save its members the trouble of forming their own character"[98]. Compression is hurtful in science not only to geniuses, but to those other talented people who differ from them, as we have seen (p. 37), not in kind but only in degree.

The danger of the threat to individualism has already been recognized in the arts. Sackville West writes[143] that he hopes that "ill-health or other extraordinary circumstances" will keep some composers apart from the unified stream of contemporary life, so that there may be Chopins and Brahmses in the future. One agrees that ill-health is a small price to pay for that; but it is a striking commentary on present-day life that the mania for uniformity should create a wish for someone's ill-health, so that he can develop his own personality and produce good music. For myself, I heartily wish all my colleagues (and myself) ill-health, if that is the necessary price for permission to live and serve the community as individuals.

Every scientist benefits from the work of others, either by personal contact or by reading. The contact between scientists is necessarily somewhat closer than between artists or composers: it is more important for them to keep in touch to some extent with what others are doing (though when this is overdone, originality wanes). Few things in life are so stimulating and productive as discussions between scientific colleagues. Sometimes two or three will find it helpful to work together for a time. This is a very different thing from dictated team-research. Anyone who has had long experience of undergraduates in a honour school of science knows that they may be roughly classified in two groups. One lot—the majority—are not fertile in ideas, and when they have got their degrees they are happy to take their places in a team, and will do very useful work in it. These men are often talented, and it would be far from my intention to try to belittle them or pretend that they do not make contributions to knowledge. The other lot consists of the born investigators. One sees them from their early days in the university, bubbling over with ideas for research, in trouble rather from the profusion of attractive subjects for investigation than from incapacity to think of anything to study. To talk to them is to be stimulated. To force them into a team or to make them follow slavishly someone else's ideas would be like harnessing race-horses together to pull a plough, or taking fighter-pilots from their aeroplanes to man a cargo-vessel. It is absurd to pretend that these men are selfish because they want to follow their own bent; other scientists are in their debt for the stimulus they impart to all with whom they come in contact. They embody the living spirit of discovery. Let them work by themselves, or collaborate for a time with equals, or draw a group of fresh young minds round themselves if they wish; but let them be free, for freedom is necessary for the development of their talent.

It would be unnecessary to say all this, if there were more understanding of the human nature of the born investigator; but alas! the general average of mankind, as J. S. Mill wrote[98], "are not only moderate in intellect, but also moderate in inclinations: they have no tastes or wishes strong enough to incline them to do anything unusual and they consequently do not understand those who have". If science is to survive, it is essential for people to understand that the real investigator must have freedom or his genius will wither.

People have tried[89] to equate individualism with the survival of the fittest, as though the only original thing an individual could think of doing were to gain an advantage over someone else. What a travesty of the facts! Listening to the work of a great composer, looking at a great picture, reading a great piece of literature, studying the results obtained by a great scientist, one is sincerely grateful that the person to whom one stands in such irredeemable debt was able to develop his own individual personality and was not forced by the severities of convention to stand at the same dull average level as the community as a whole. The modern tendency to confound individualism with selfishness, and collectivism with generosity, would be funny if it were not a real danger to civilization. Nothing is more selfish than finding happiness in the direction of the lives of others, nothing more generous than the wish that all should be free. Let those who pretend that individualism is selfish ask themselves whether they find generosity in the concentration camps, purges and so-called "trials" of totalitarian states.

It is instructive to make a comparison between scientists and soldiers. In an army the best-informed person is supposed to be the general, who deputes duties less important than his own to colonels, and these do likewise to majors, and so on through captains and lieutenants and the various non-commissioned ranks down finally to private soldiers, who form the largest and lowest group of all. In the world of scientific research we must equate the investigator with the private soldier and lance-corporal, the latter rank only appearing when a few scientists agree to work together with one of their number as the main animator of the investigation. (I leave out the laboratory assistants, not because I underrate their great services, but because they are generally fewer than the research workers and thus confuse the simile.) If research were dictated from above, there would have to be higher ranks, all the way up to generals. This would lead to absurdity. No born investigator would ever agree to rise above the rank of lance-corporal and thus voluntarily become a petty little scientific dictator while depriving himself of the one thing he values above all else in life, the actual doing of research. All the higher ranks would have to be filled, therefore, by people who are not first-rate research workers. To imagine that the true investigator would work under their direction is equivalent to imagining that a composer would agree to put his notes where a less talented person thought fit. The whole idea of dictated research is fantastic, because the people best qualified to be generals would insist upon being lance-corporals. There can be no hierarchy in research.

So long as science contains only lance-corporals and privates, there will be no interference with research. Karl Darrow, of the Bell Telephone Laboratories, has put this matter succinctly[32]. He says that the essential requirement for the development of physics in particular and science in general is "a supply of talented people enabled and permitted to go their own ways, so that discovery may occur in whatever logical or capricious or ironical ways may be chosen by destiny. Non-interference is essential". It is essential because discoverers must be free gradually to develop their talents and move from topic to topic and even from subject to subject as their experience develops.

The development of experience can be illustrated by the life of Pasteur, who never suffered the indignity of having the subjects of his research dictated to him by others, but moved freely forwards through a wide series of investigations, each suggested by the one that went before. Starting with the study of the forms of the crystals of tartrates, he finds that one tartrate will undergo fermentation if infected with albuminous matter, while another, differing only as left from right, will not. He leaves crystal structure, having made an enormous contribution to that study, and passes to the investigation of fermentation, at that time so little explored. He works at the fermentation of milk and the resulting formation of lactic acid, and makes the illuminating discovery that bacteria are the cause of the chemical change. He passes to the question whether bacteria can originate from inanimate matter or only from pre-existing bacteria, and draws the latter conclusion. From this momentous work he passes to severely practical issues, the rôle of micro-organisms in the vinegar and wine industries, and afterwards from the diseases of wines to those of organisms. Having shown that micro-organisms can cause disease in silkworms, he passes to methods of counteracting their harmful effects. He discovers that the germ of chicken cholera can be rendered less virulent, while retaining its capacity to give immunity to fowls against the virulent form of the disease.

He applies this method to the prevention of anthrax in sheep and cattle, and passes lastly to his crowning practical achievement, the conquest of hydrophobia. So, by a round-about route, Pasteur passed from crystallography to immunology, and from one epoch-making discovery to another. Never did anyone try to confine his attention to set subjects.

Let us compare Pasteur's freedom with the shackles surrounding the research worker in a totalitarian state. In Soviet Russia[26] a research worker cannot change his subject without wide discussions with the rest of the staff of his institute, and an individual's personal desires as to what he wants to work at receive little consideration. Research workers are organized in brigades. There is little cause for surprise when even a convinced admirer of the Soviet régime has to admit that "it would be idle to look for the quietly pursued excellence and sound and acute scholarship" in the U.S.S.R.[16]. Bernal may make what excuses he likes for his protégés, but it will not alter the fact that quietly pursued excellence and sound and acute scholarship are precisely what we should look for everywhere in science. If scientists of the future are everywhere organized in brigades, and little consideration paid to the desires of individuals as to what they want to study, then genuine investigators will fit up primitive little laboratories in attics and sheds, and the great discoveries of the future will be made at home.

It is a striking fact that even in communities where the supposed welfare of the state is made the supreme consideration, the authorities dare not ignore the existence of the individual. When Maria Utkina, unaided, destroys a Panzer formation or performs some other commendable military exercise, we are told very particularly in the Soviet communiqué that it is Maria Utkina herself who did it. There is an inherent recognition of the fact that even in a totalitarian state, potential Maria Utkinas appreciate the performance of an individual.

The forcing of research workers into brigades is a deprivation of their rights as individuals, but an even more intolerable proceeding would be the amalgamation of all the scientists in a country into a single body intended to get power by the threat of striking. The question of the advisability of a strike by all the scientists in our country is seriously discussed by J. G. Crowther[26], who does not realize that no section of the community is less interested in possessing power than the genuine investigator. One has only to think of the great scientists of the past and present and of the stimulating personalities among one's own scientific friends and colleagues, to realize that petty struggles for power would be beneath their contempt; and they would consider it nothing short of idiocy voluntarily to deprive themselves by striking of the thing above all that makes their lives seem to them worth living, the right and duty to discover truth by free enquiry.

Free enquiry is just as important to the scientist as freedom of speech and publication. This point is so obvious that it might be thought scarcely worth mentioning, but in fact it is very important to mention it. Many people who talk airily about the necessity for freedom in science are actually thinking only of the freedom of scientists to say and write what they like. That is a relatively small freedom, if they are not allowed to investigate what they like and therefore to make discoveries which are worth talking and writing about. There is a movement on foot to try to get university laboratories to undertake research on stated problems after the war. This is a real threat to freedom. A biologist might wish and feel himself fitted to devote himself to research on, e.g., cell-chemistry, but his university department might be allotted the task of discovering how to induce the larvæ of clothes moths to make fewer or smaller holes in clothes. It would be no good to say to that man, "My dear sir, you are perfectly free: you can publish whatever you like on cell-chemistry or politics or anything else." The man would answer, "I want to publish on cell-chemistry, if I can make discoveries in that subject, as I have reason to believe I can. If I may not work on that subject, my freedom to publish about it is illusory." The negation of freedom is implicit in the suggestion that science should be centrally planned.

On 28th September 1941 the President of the British Association announced at the Royal Institution, London, a "Charter of Scientific Fellowship". The second word is freedom and the necessity for liberty is re-iterated, but it is clear that those who drafted the Charter had freedom of speech and publication, and not freedom of investigation, in mind. The people who wish to see science planned seem to have had such an unfortunate influence on those who drafted the Charter, that its preamble and seven clauses contain nothing to suggest that any scientist anywhere should have the right to decide what he will investigate. Freedom of thought and its expression serves the philosopher and writer well: it is not sufficient for the scientist, who wants to find out first and talk afterwards. True freedom is not granted by this charter, but it is a cardinal part of the requirements for liberty laid down by the Society for Freedom in Science. One of the five propositions to which members of that Society adhere is this: "The conditions of appointment of research workers at universities should give them freedom to choose their own problems within their subjects and to work separately or in collaboration as they may prefer. Controlled teamwork, essential for some problems, is out of place in others. Some people work best singly, others in teams, and provision should be made for

D

both types." These few unequivocal words are worth more than any amount of loose talk about freedom.

If the community were to decide that all research workers must contribute directly and obviously to material human welfare, they could be required to spend part of their time on manual or other labour. Already the majority of research workers spend a large part of their time in teaching or in other work in which complete freedom is impossible. (Lecturers, for example, must give their lectures on appointed subjects at appointed times.) If this were not considered a sufficient direct contribution to material welfare, the community would be far better advised to make research workers do a certain amount of manual labour and leave them free in their research periods, than to conscribe them into planned research.

In special research institutions it is reasonable to require people to undertake the sort of research for which the institution was founded. For instance, a scientist who has freely joined a marine biological laboratory may properly be required to devote his attention mainly to research in marine biology. When a research worker does no teaching or consultation but devotes the whole of his time to research, it is not unreasonable to require him to spend a part of his time in working on a subject selected by the director of the institution, provided that he is free during a substantial part of his time. With these and similar exceptions the utmost freedom is desirable, and it scarcely needs to be said that under all circumstances the investigator should be free to publish his results, whatever they may be.

It is fashionable to set up an Aunt Sally of individualism in order to knock it down. When great power is placed in the hands of single persons, they often use it to the detriment of others; and a baseless hatred of individualism has grown up because of this and as a result of envy and jealousy.

Those who would use their rights as individuals to the greatest benefit of the community as a whole—the composers, artists, writers and scientists—are the very people to whom power means least and who are the least likely to wish to hold others in subjection. To deny individualism to the creators of culture would be to make their work impossible; to do so on the ground that lovers of power misuse it would be absurd. Creative thinkers do not desire to control but only to serve humanity. It is vital to civilization that true individualism should survive and be saved from confusion with the personal love of power, with which it has nothing in common. "Whatever crushes individuality", a great philosopher has written[98], "is despotism, by whatever name it may be called."

Those who care nothing for freedom try to confuse the issue by saying that the urgent freedoms are freedom from want and freedom from fear. This obvious and self-condemned attempt to evade the issue should deceive no one. Freedom from want and fear is intended to mean absence of want and fear. Everyone desires an absence of what is obviously undesirable. Freedom means nothing negative. The word implies the positive right of individuals to choose for themselves between different possibilities of action.

Chapter 4

Planning in Science

"Le hasard ne favorise que les esprits prépares." PASTEUR[136]

WHEN it is proposed to build a house, a plan is drawn. That is because it is possible and necessary to envisage the final form of the house before building it. It is precisely because the whole purpose of science is to discover that which is not and cannot be envisaged, that planning in science is self-contradictory; it is as though explorers were to map an unknown country before they had reached it.

Today there is a resurgence of an old idea that almost anyone can be machine-finished into a scientist and that advances in knowledge can be turned out automatically like sausages from a machine. Those who want to plan science go back to the sixteenth and seventeenth centuries to find their hero in Francis Bacon, than whom a more unsuitable object for heroics can scarcely be conceived. Those who see in him the prototype of that modern repository of virtue, the politically-minded planner of science, should know that Bacon's interest in politics was fostered solely by his desire to gain worldly power. He obtained the personal advancement he so much desired by concealment of his own opinions and by flattery. When it suited his convenience, and power was in his hands, he turned savagely against those who had befriended him. It is strange indeed that those who think themselves revolutionary in politics should admire Bacon. When an old clergyman wrote a sermon justifying insurrection under certain circumstances, Bacon was active in his prosecution, though the matter was never preached or published. He questioned the old man under torture, and visited each judge of the court separately in private, to secure conviction. One so obsequious to the king as Bacon is a queer hero for modern times. His conviction for taking bribes to influence the course of justice in lawsuits ended his public career.

When all this has been said, it remains true that Bacon could be respected if his scientific work were respectable. It has been pointed out, however, by a historian of physics[20], that he "was not a scientific man; he had little practical experience in experimentation; he lacked the scientific instinct to pursue in detail the great truth that nature must be studied directly by observation and experiment". Bacon thought he could make almost anyone into a scientist, for his method "nearly levels all wits and intellects"; he would have been surprised if he had been told that three centuries later a scientist would write of his method that "no evidence can be shown of its successful application in any branch of science"[19]. Draper[39] says of him very simply that he was "a treacherous friend, a bad man", whose admirers have thought that scientific discoveries are accomplished by a mechanico-mental operation. Oliver Lodge[88] said that "on the solid progress of science he may be said to have had little or no effect". Andrade[6] says that he was "a stranger to quantitative work, and he had an aversion from the method of the working hypothesis, to which science actually owes its advances . . . he could not recognize real scientific advances when he saw them".

Bacon's writings contain a mass of ill-digested and disconnected statements about natural objects and phenomena, with suggestions for experiments but little evidence of a desire to try them himself. His "Sylva Sylvarum"[137] is scarcely readable, so full is it of uncritical statements, errors, capricious formulae and platitudes. "A *Man Leapeth* better with *Weights*, in his *Hands*, than without. The *Cause* is, for that the *Weight*, (if it be proportionable,) strengtheneth the *Sinewes*, by *Contracting* them." "A *Dry March*, and a *Dry May*, portend a *Wholsome Summer*, if there be a *Showring April* between: But otherwise, it is a *Signe* of a *Pestilential Year*." "It is true, that the *Ape* is a Merry and Bold *Beast*. And that the same *Heart* likewise of an *Ape* applyed to the *Neck*, or *Head*, helpeth the *Wit*; And is good for the *Falling-Sicknesse*." It is strange to reflect that this book was written by one who ignored Harvey and Kepler, rejected Copernicus, and did not appreciate the achievements of Gilbert and Galileo. It is stranger still that some people should derive the methods of modern science from him rather than from them. His guess about the nature of heat was not a discovery in any scientific sense, and anyhow a very similar guess had been put forward by Plutarch more than a millennium before[144].

The idea of planned research originated in Bacon's "New Atlantis"[138]. This is an imaginary story of the author's visit to an unknown island called Bensalem in the Pacific Ocean. His ship had sailed from Peru for China and Japan, but was delayed by contrary winds. The ship's company ran short of food and had given themselves up for lost, when they sighted an island. Here they were most hospitably received by the inhabitants, who had had no visitor from the outside world for thirty-seven years, and very few for many centuries past. In ancient days the island was known to the other parts of the world, but about 300 B.C. the great king of Bensalem, Salomona, decided that contact with strange lands should be almost completely avoided, as he thought it difficult to derive benefit from receiving strangers. Twenty years after the death of Christ an enormous apparition of God's finger converted all the inhabitants to Christianity, and a cedar box was conveniently sent, containing the whole of the Bible, including the parts which had not then been written. The people became extraordinarily virtuous in all respects and remained so down to the time of the author's visit.

The author was chosen from the whole of the ship's company for the very special privilege of visiting the head or "Father" of a gigantic technical college called Salomon's House, which had been instituted by the great king who first caused the island to be isolated. Salomon's House extended vertically from towers half a mile high to caves three miles below the surface of the earth. Its charter was a worthy and dignified one: "*The end of our* Foundation is the *Knowledge of* Causes, *and Secret Motions of things; and the Enlarging of the bounds of* Humane Empire, *to the Effecting of all things possible.*" In this college the principles of planned technology were carried into effect. Little was done in pure science, whose study was confined to three "Mystery-men". There were "Particular Pools" where "Trials *upon* Fishes" were made, and "Perspective-Houses" for optical experiments; but beyond this most of the work was in technology. After meetings and consultations of all the investigators, three special Fellows devised new experiments for others to undertake. Each Fellow had his job: some were professional students of the literature of the subjects investigated at the college; others were explorers, charged to reside abroad for twelve years at a time, concealing their place of origin, and to bring back information about foreign technology and sciences. The experiments were concerned with remarkable enquiries. Some applied themselves to perfect the water of Paradise, "*very soveraign for* Health *and* Prolongation *of* Life". Others concocted "Drinks *of* Extreme Thin Parts; *To insinuate into the* Body". Others again worked to "Multiply Smells", which, as the author justly remarks, *"may seem strange"*. Yet others *"represent all manner of* Feats *of* Jugling". Among their laboratory conveniences were "Heats *of* Dungs, *and of* Bellies *and* Mawes *of* Living Creatures".

The story ends abruptly when the author had been told about the marvels of the Academy, but had not himself seen them; for "New Atlantis" was never finished.

Salomon's House lends itself to satire, and it is probable that Swift's Academy of Lagado[124] is a skit upon Bacon's idea. It will be recollected that when Gulliver is let down from the flying island of Laputa and reaches the earth near the city of Lagado, it is not long before he is taken to see the wonders of the Academy. It is amusing to read the "New Atlantis" first, and then turn straight to Chapters 5 and 6 of the Laputa voyage, in which the visit to the Academy is described. Swift is here in his most satirical vein. The modern exponent of the elaborate card-index and other mechanical devices for enabling anyone to make discoveries automatically finds himself already satirized long before his birth. A professor of the academy tells Gulliver that everyone knows "how laborious the usual method is of attaining to arts and sciences"; and he explains his mechanical contrivance whereby "the most ignorant person, at a reasonable charge, and with a little bodily labour" can write books on any subject "without the least assistance from genius or study". Even Swift's vivid imagination scarcely sufficed to make the employments of the professors at the academy more fantastic than those which occupied the attention of some of the Fellows of Salomon's House.

Bacon was no scientist and did not understand the conditions under which research can be carried out. Nevertheless, there are a few things that stand directly to his credit. He strove to overthrow belief in the ultimate authority of Aristotle. He was far before his time in his attitude to teleology. In "The advancement of learning"[9] he shows that little help is given by statements that the eye-lashes are intended to protect the eyes, or skins and hides to defend from heat and cold, or bones to form frames for living creatures, or clouds to water the earth. The search for physical causes, said Bacon, "hath been neglected, and passed in silence", because the teleological explanations have acted as hindrances. This is sound sense, if not pushed too far. Better still is one inconspicuous remark in "New Atlantis". When describing the functions of the various Fellows of Salomon's House, he tells of three who differ markedly from all the others. In the words of the Father of the House: *"We have Three that trie New Experiments. Such as themselves think good. These we call Pioneers or Miners."* Much can be forgiven to Bacon for those eighteen words. Science was not quite dead even in the metropolis of planned technology.

The fundamental reasons why science cannot be comprehensively planned are two. First, scientists are not, and cannot be replaced by, crude machines, a fact that has been sufficiently emphasized in the foregoing chapters. Secondly, discoveries often come by way of surprise, and not as a result of meticulous attention to a plan; for the end of a research cannot be envisaged when the plan is being made. If a scientist is working along a line of his own, and suddenly an unexpected vista opens before him, he will often arrive at important results through having the necessary insight and mental energy to throw over his plan and follow the new trail[10]. Chance plays a large part in discovery, much larger than people are apt to think; but chance, as Pasteur said, "only favours prepared minds"[136] It is not the slave to a plan, but the person with the mind that is prepared for the unexpected, who becomes the great discoverer.

The discovery of current electricity was made by that

fruitful combination, chance and a prepared mind. Galvani was not a physicist, but a professor of anatomy at Bologna. In his experiments on muscles removed from recently-killed frogs, he sometimes noticed unaccountable twitchings, not attributable to any obvious mechanical stimulus. He experimented for several years on frogs' legs suspended on wires and hooks, and at last discovered that the twitchings were caused by the accidental touching of one part of the preparation by a piece of iron and another by a piece of copper, the iron and copper being themselves in contact.* Galvani had seen similar twitchings produced by electrical machines, and he thought there might be an electrical explanation[2]. It was true: he had accidentally made a primitive electrical cell. His observation led to the understanding of current electricity and to the invention of the voltaic cell by Volta.

Von Röntgen had no thought of trying to make human flesh transparent when he discovered the penetrating powers of X-rays. He was interested in the phenomena of electric discharge in high vacua, and did not guess that the result of his work would be the discovery that certain rays could be used in the diagnosis and treatment of human illness. As he afterwards said himself, "I found by accident that the rays penetrated black paper." He had been using barium platinocyanide with the object of detecting invisible rays, but with no thought of such rays being particularly penetrating. He left some of this substance on a bench near his vacuum tube, and chanced to notice that although the room was dark and there was black paper in between, the barium platinocyanide became fluorescent[10, 52]. From the time of seeing the fluorescence von Röntgen isolated himself for three weeks, working at high pressure without discussing the phenomenon that he had witnessed with anyone. At

* Another explanation of Galvani's results is possible.[146]

first he ate and slept in the laboratory. As Gregory has written[55], "After the discovery of Röntgen rays, their application to medicine was soon seen." "After" is the key word.

Von Röntgen's discovery displays two quite separate aspects of the action of chance in science, for it illustrates unexpectedness both in discovery and in the applications of discovery. In the first place, the original discovery in pure science—the discovery that the rays are particularly penetrating—was made quite by chance, owing to some barium platinocyanide happening to be in the vicinity of the vacuum tube. In the second place, it was by chance and not by plan that medicine was so enormously benefited. If someone had thought it would be convenient to make the human body transparent, and had allocated money for the research, the result would have been a comprehensive plan, a team of research workers, a very large card-index, a waste of money, and no X-rays. Von Röntgen wanted no team and no planner's plan. He wanted to make his own studies of electric discharge in high vacua and he wanted to work alone. Not only did he not want a single colleague in his laboratory: he did not even want a laboratory assistant unless a piece of apparatus required the use of more than two hands, or his green-colourblindness necessitated outside help. He preferred the undisturbed atmosphere of solitude during his research[52]. Not many men have done more for pure science or for the relief of suffering than von Röntgen.

In the pages of this chapter and the next a few instances out of a multitude are given of the action of chance in science. One might have given examples of its action in discovery first, and then passed on to examples of the unexpectedness of the applications of discoveries to practical affairs, but it has seemed best to consider both kinds of chance together. Often the two kinds are woven together, as in the discovery and application of X-rays and of the relation between electricity and magnetism.

It was the happy combination of chance and a prepared mind that disclosed the relation between electricity and magnetism in 1822. Because a magnetic needle chanced to lie near a wire carrying an electric current, Oersted stumbled on his epoch-making discovery. It was of this particular event that Pasteur made his remark about chance favouring the prepared mind. The converse discovery by Faraday, that a magnet can cause an electric current to flow, illustrates with overwhelming force the unexpectedness of the applications of science. If we ask ourselves what scientific discovery has, more than any other, changed the face of civilized life in a material way, we are likely to answer the discovery of electro-magnetic induction. The basic fact that the movement of a magnet in the vicinity of a coil of wire produces an electric current was made by Faraday as a contribution to pure science. The electrical, motor-car and aeroplane industries depend for their existence upon that discovery. Its huge applications render all the more amusing the memorable incident which occurred when Faraday had just delivered a public lecture on the subject. A woman in the audience asked Faraday what was the use of his discovery. If she could live today and see what have been its scientific and practical results, she would blush with shame every time she remembered the incident. In this age, when real science is threatened by those who vainly pretend that discoveries come only in direct response to human needs, it is exhilarating to recall Faraday's answer, a repetition of what Franklin had said in similar circumstances long before[136]. "Madam," asked Faraday, "will you tell me the use of a new-born child?"[55] The history of electromagnetic induction exposes as clearly as X-rays the lie that necessity is the mother of invention.

Gregory[55] has told how the readers of an American magazine were asked to choose by vote which were the seven most marvellous discoveries and inventions of the modern world. The readers chose wireless telegraphy, the telephone, the aeroplane, radium, antiseptics and antitoxins, spectrum analysis, and X-rays. As Gregory pointed out, each of these discoveries and inventions had its foundations in pure science and was not the result of deliberate intention to make something useful.

The history of the microscope shows convincingly how love of knowledge, quite apart from practical applications, is the driving force behind the progress of science. Anyone who follows the evolution of the compound microscope from its invention by Janssen in 1590[38a] to the magnificent instrument of modern times knows that all along progress has been due to the desire to see the ultimate detail of small objects more clearly for the sake of increasing knowledge. The turning of the instrument to use in industry followed far behind its development for use in pure science. One of the landmarks in its history was the discovery by a wine-merchant[53], Joseph Jackson Lister, that two separate compound lenses could be put one behind the other to form a new sort of high-power object-glass in which the error called "spherical aberration" was largely abolished. Lister was an amateur microscopist, yet his invention ranks in importance with Dollond's discovery of how to minimize colour-fringes and Abbe's magnificent achievement in almost entirely eliminating them by means of his "apochromatic" object-glasses[115]. Abbe's heart and soul were in the improvement of the microscope. The motive of commercial profit did not count with him, and he generously renounced the ownership of his optical works in favour of his employees. Students of the minute plants called Diatoms were always anxious to get the best possible instruments so as to see the striations and dots on their siliceous shells that are the ultimate test of the resolving power of microscopes. Computers and manufacturers strove always towards perfection: the "Diatom-dotters" would be satisfied with nothing short of the best. The Managing Director of the largest firm of microscope manufacturers in Britain has said[141] that the greater part of the evolution of the modern microscope is to be attributed to the Diatom-dotter. A combination of inventors like Lister and Abbe with enthusiastic observers of microscopic life produced, through love of the subject, the wonderful instrument that we know today. That instrument is at the service of industry, and its material value can scarcely be exaggerated; but the many industries which now profit from its use did nothing to help its development.

Chapter 5

Planning in Science

continued

In the preceding chapter a few instances have been quoted to show how important the rôle of the unexpected has been in physics, both in discovery and in the applications of discovery. Innumerable instances could be chosen from other sciences. The discoveries of aniline by Unverdorben, of benzene by Faraday and of chloroform by Soubeiran were instanced, among others, by Crookes as the work of men who "seemed at the time never likely to be of the slightest use to anybody"[25]. Such men revolutionize the material welfare of society, while those who follow a plan to improve what already exists plod slowly forward. Science progresses by work in fields whose applications cannot possibly be envisaged. When Wollaston discovered the elements palladium and rhodium, how could he have guessed that the one would find practical application in toning solutions used in photography and the other as a platinum alloy in thermocouples, both photography and thermocouples being concepts of a later period?

People who are not themselves connected with scientific research would be apt to imagine that a substitute for sugar would be discovered by a research undertaken for the purpose of finding a substitute for sugar. That is not how things happen, or can happen, in science. The discovery of saccharin provides an outstanding example of the function of chance in discovery. Remsen and Fahlberg[112] had no thought of discovering a substitute for sugar when they made their investigations. Their object was purely academic. They wanted to know whether it was possible to make orthosulpho-benzoic acid from orthotoluenesulphonic acid. Before this was done, a successful research by another worker on a closely related subject caused their project to lose much of its importance and interest, but they decided to persevere and produced what they called benzoic sulfinide. There was nothing to suggest that their research would have practical applications, until someone tasted the product. "It possesses *a very marked sweet taste*", they wrote, "*being much sweeter than cane-sugar*. The taste is perfectly pure. The minutest quantity of the substance, a bit of its powder scarcely visible, if placed upon the tip of the tongue, causes a sensation of pleasant sweetness throughout the entire cavity of the mouth." They had discovered saccharin. The present-day product differs only in the addition of another substance which, though itself less sweet than saccharin, increases the sweetness of that substance still further[131]. Perfectly pure saccharin is already five hundred times as sweet as sugar, and Remsen and Fahlberg did not exaggerate when they marked their surprise by the italics given in the passage quoted above.

Chemistry provides many examples of the action of chance, but none perhaps more striking than that on which the synthetic dye industry is based. W. H. Perkin, senior, when only eighteen years old, tried to produce quinine synthetically by the oxidation of allyl-*o*-toluidine by potassium dichromate[64]. He failed, but thought it might be interesting to find what happened when a simpler base was treated with the same oxidizer. He chose aniline sulphate and obtained a black precipitate. This included a purplish substance, which was found capable of imparting its colour to silk. Perkin had discovered the first aniline dye. The whole of the vast synthetic dye industry was built on that discovery; and Perkin was not investigating anything connected with colour when he made it. Chance played an even bigger part in the discovery than the bare facts might indicate, for he would never have discovered aniline dyes if his aniline had been pure. By good fortune it contained some *p*-toluidine, without which the reaction could not have occurred.

A little more than a century ago, when Daguerre was making his attempts to fix the visual image which could so easily be thrown on a screen by a lens, he tried using iodized silver plates, because he knew that silver iodide was sensitive to light. He discovered accidentally[95] that the exposure of his iodized silver plates to the fumes of heated mercury produced a light-sensitive material which gave a fixed image of the object focussed by the lens. Thus the first kind of photography was born—by accident.

Chance often starts a scientist on an investigation which he would not otherwise have undertaken. It appears that the first catalogue of the stars owed its existence to the sudden appearance of a new star during the second century before Christ. This event is said to have influenced Hipparchus, the famous Greek astronomer and founder of trigonometry, to make a list giving the positions of over a thousand stars[54].

Chance started Darwin on his great study of insectivorous plants. In the summer of 1860 he was idling and resting near Hartfield in Sussex[36]. His great work, "The origin of species", had been published the year before, and his mind was ready for a new adventure. It happened that in the neighbourhood of Hartfield two species of sundew (*Drosera*) were abundant. It will be recollected that the leaves of this plant catch and digest small flies. Darwin's attention was instantly attracted: his enthusiasm has been recorded on a previous page (p. 18). The chance presence of sundews

E

at the place where he took his holiday resulted fifteen years later in the publication of his book, "Insectivorous Plants"[34].

That work of Darwin's was of great interest in pure biology, but did not lead to practical applications. Another botanical study of his, also undertaken purely for its scientific interest, has been fruitful materially as well as academically. The tracing of this subject on its long and unexpected journey from the laboratory of pure science to commercial exploitation shows how unpredictable the course of discovery is, and how destructive it would be to tie research workers down within the narrow confines of a plan. We owe the valuable chemical root-stimulants of today to the research of workers in pure science who were investigating a problem wholly unconnected with the stimulation of root-growth. The subject of their studies was the bending of plants in response to light.

Charles Darwin[36] found that it is the tip of a shoot that perceives the light, but that lower regions, which have not themselves perceived it, bend nevertheless in response. An influence of some kind clearly passes downwards from the tip. In 1913 Boysen Jensen showed than an influence of this kind could pass through plant tissues which he had completely cut across and which were then fixed together with gelatine. This suggested that a substance must pass from the tip through the gelatine to the tissues lower down, which reacted by bending. It was Paál who proved this. Both Boysen Jensen and Paál worked with the seedlings of grasses, the former with the oat, *Avena*, which became the classical object for the study. The young shoot of the germinating seedling is protected by a long, hollow, conical or pencil-shaped sheath. Paál cut this transversely across, as did Boysen Jensen, and then stuck it on again; but instead of sticking it on centrally, as one might mend a pencil broken in two, he put the detached part back somewhat to one side, so that its base only covered part of the exposed stump.

Now a very remarkable thing happened. An influence passed down from the tip of the detached part into the tissues of the seedling, below where the cut had been made. The influence caused those tissues to grow fast, but it only affected that side above which the cut-off part was re-attached. The result was that growth was greater on one side than on the other, and the shoot began to bend. Paál rightly concluded that a growth-stimulating substance had passed down one side only of the seedling and thus produced asymmetrical growth.

Thus arose the study of plant-hormones[100, 142]. Chemical messengers of this kind, originating in one part of the organism and affecting the growth of another, had long been known in animals: the thyroid, pituitary and reproductive glands are among those which produce them and thus exert powerful influences on growth. In plants they had hitherto been unknown. Now the problem was to identify them. Went proved in 1928 that the substance formed by the sheath-tip, whatever it (or they) might be, was unaffected by boiling and therefore not a ferment. Others sought to find out what substances other than those which originate in the shoot-sheath can permeate plant tissues and influence growth. Strangely enough it was found that human urine contained large quantities of two such substances, "auxin a" and a substance called indole-acetic acid. It would appear that "auxin a" is the main hormone actually produced by the shoot-sheath of the oat, and it seems very extraordinary that it should occur also in urine. Indole-acetic acid was shown to be the very same substance that a certain fungus makes, and it greatly affects the growth of the accepted test-object, the seedling of the oat.

Various people have from time to time tried deliberately, but with little success, to stimulate root-growth, mostly by experimenting on the nutrition of the root. Nutrition would be the obvious subject of the planner's plan, but it was not the solution of the riddle. Bouillene and Went showed in 1933 that the first leaves of seedlings contain a hormone which stimulates root-growth. Substances with the same effect were extracted from various plants, especially the fungus already mentioned. In 1935 Thimann and Koepfli[128] and another independent worker showed that indole-acetic acid, prepared in the laboratory, is as effective as any of the natural products.

The responses of the roots of many plants were now studied. Practical applications were already envisaged in the year of Thimann and Koepfli's discovery, by a worker who tried root-stimulants on plants of commercial importance. Favourable results were obtained and the method has been taken into horticultural practice. The roots of seedlings are treated for a day or less with a solution of indole-acetic acid in water, and a surprisingly prolific outgrowth of rootlets subsequently results. Cuttings may be stimulated to form roots by being smeared with lanolin mixed with a little indole-acetic acid. Thus purely academic investigations of the influence of light on the movements of plants have not only greatly enriched pure science, but also, in the most curiously roundabout manner, resulted in the introduction of a method of direct material advantage to man. Went and Thimann, two of the foremost workers on plant hormones, have themselves pointed out that it was research in pure science that gave the world the practical benefit of root-stimulation[142].

Endless examples of the effect of the unexpected in the promotion and application of discoveries could be quoted from any science. I shall limit myself to a few more, mostly chosen from the sciences on which medicine is based.

More than half a century ago von Mering and Minkowski were studying the function of the pancreas in digestion. In the course of their work they removed the pancreas from a number of dogs. By an extraordinary bit of good fortune, a laboratory assistant happened to notice something unusual about those dogs: swarms of flies gathered round their urine[23]. Today, thousands of people have reason to be grateful to that laboratory assistant for mentioning it to the investigators. One of them thought it worth while to look into the matter. The urine was analysed and found to be loaded with sugar. Experimental diabetes had been produced for the first time, and an insight given into the cause of that disease. It was not some one working at diabetes who found the cause. The investigators were studying the physiology of digestion, a subject which has no connection with diabetes. From then onwards it was known that the pancreas produced something as well as digestive ferments. That something was insulin, and the problem was how to isolate it. If that could be done, the substance could be injected into people suffering from diabetes. The difficulties were great, because the digestive ferments of the pancreas destroyed the insulin when attempts were made to extract the latter. Banting and Best achieved the first solution of the difficulty[91]. They tied off the pancreatic duct of animals and thus caused the cells which produce the digestive ferments to atrophy, leaving the insulin unaffected and extractable. It would be too troublesome, however, to have to make a major operation on an animal every time one wanted to use its pancreas as a source of insulin, and then wait for changes to happen in that pancreas. With the aid of Collip (see p. 28), they succeeded in making extracts of insulin from ordinary pancreases without any necessity to tie the duct, extracts so pure as to cause no local irritation when injected into human beings. Insulin enables diabetics to control their illness.

The long researches which led to insulin started from a chance observation. Again and again in science it is a chance observation or even a mistaken idea which leads to great discoveries, not solemn plodding along the dull paths of some one else's pompous and consistent plan. The whole subject of chemotheraphy, with all its enormous benefits to the human race, was founded on a mistake. Paul Ehrlich was interested in the action of dyes in rendering distinct the minute structure of the tissues of animals. In 1886 he gave the formula[42] for a dye solution which is probably to this day more used in biological laboratories than any other. He reflected on the specificity of certain dyes for certain micro-organisms, and thought it possible that a substance might be found which would combine with the protop asm of the living parasites causing disease without damaging the host tissues. He tried the effects of various dyes on living trypa-nosomes, those blood-parasites of which some cause sleeping sickness in man and other nagana in cattle. In 1906 he found that trypan violet was effective in the way he had hoped[127]. This discovery was the origin of chemotherapy. The dye was itself too harmful to the host to provide a solution to the problem, but it was more harmful to the parasites. Ehrlich now tried a different but related kind of drug, differing mainly in that two atoms of arsenic replaced two of nitrogen in the central part of the molecule. The result was "606", which, with the related drugs that followed it, has been so enormously effective against syphilis. Thanks to them, it is not easy today to find a fully-developed case of the disease to show to medical students. Sleeping sickness fell to trypar-samide, and the diseases due to trypanosomes and spiro-chaetes were no longer the menace they had been. This huge success was founded on an error, for Ehrlich's idea, that the drugs would act on micro-organisms as a dye on wool, was wrong. Various theories of their mode of action are held today, but Ehrlich's is not among them[127]. A very great man, following a false clue, became one of humanity's greatest benefactors.

The organic arsenical drugs which owe their existence to Ehrlich are not effective against the diseases caused by bacteria, and blood poisoning, puerperal fever, gonorrhoea, etc., remained the scourge they had always been. Today those diseases are under man's control, thanks to the following up of another false clue by another great man, Gerard Domagk.

Domagk was impressed by Ehrlich's early work, and tried the effects of a great number of dyes belonging to the group called "azo-dyes", to which Ehrlich's trypan violet belonged. It was in 1932 that the great discovery was made. A drug was found which killed streptococci without killing the mice they parasitized. In 1935 the azo-dye, prontosil, was put at the disposal of medical men all over the world and a new era in medicine was opened, the era of the attack on bacteria by chemotherapy. Puerperal fever, blood-poisoning, erysipelas, pneumonia, cerebrospinal meningitis and gonorrhoea began to fall before the series of drugs related to prontosil. Yet the clue to this great discovery was false.

Domagk worked with azo-dyes, which are complicated chemical compounds characterized by the presence of two nitrogen atoms in the middle of the molecule. It is the presence of these two nitrogen atoms, linked together and to the rest of the molecule in a special way, that confers colour on the azo-dyes. Prontosil itself is red. Remove the azo-linkage from an azo dye, and colour is destroyed. Domagk did not break the azo-linkage, which he thought necessary for the action on bacteria. Very shortly after Domagk's work

was made public, four French workers published a little
two-page paper[134] which threw an entirely new light on the
subject. They tried the effect of one end only of Domagk's
molecule, without the azo-linkage. This end, by itself, is
sulphanilamide, a substance which is colourless and much
simpler than prontosil. They injected the hydrochloride of
this base into six mice infected with streptococci taken from
a woman who died of puerperal fever. There was a marked
effect. The five control mice, which were infected with the
streptococci in the same way but received no sulphanilamide, were all dead within forty-eight hours, while all the
six which had received the drug were still alive. The work
was soon confirmed, and today sulphanilamide is one of the
most-used drugs for streptococcal infections. It owes it use
to Domagk, for without his discovery of prontosil it would
not have been tried; but the action of sulphanilamide shows
that he himself worked on a false (though very fruitful) clue
when he made his discoveries.

Again and again in working towards practical ends one
falls back on pure science for support. The French workers
did not have to discover how to make sulphanilamide: there
was the method, ready and waiting for them. Sulphanilamide was discovered in Vienna by a man who was making a
general study of the sulphonamides of sulphanilic acid.
That man was Gelmo, and the year 1908[51]. Tucked away in
the middle of his paper is a paragraph on sulphanilamide.
He tells very shortly how he made it and gives an account of
its properties: its whiteness, solubility, reaction with alkalis,
stability to acids, and melting point (the latter 3° C. wrong,
as we now know[5]). That little paragraph is one of the most
tantalizing objects the world has ever known. Neither Gelmo
not anyone else had the faintest idea that sulphanilamide
would become one of the world's greatest life-savers.
Twenty-seven years had to pass before it was discovered

that it was a chemotherapeutic agent of the foremost rank. Taylor[127] calculates that if its properties had been known, it could have saved 750,000 lives in the last war alone, from its effects on the streptococcus of septic wounds and blood-poisoning. It was just because chemists did not short-sightedly confine their attention to substances thought likely to be useful in a material way, that sulphanilamide was lying ready for instant trial directly some one had reason to believe it might serve humanity in a practical way.

It was Pasteur who spoke of the function of chance in science while he was yet a young man, and it was fitting that he himself should derive such a signal benefit from it many years later. After he had worked for some time at chicken cholera, his experiments had to be interrupted for several weeks. That interruption, unplanned, was amazingly fertile in results. When he started work again, he reverted to his old cultures of the bacilli which cause the disease and tried to inoculate them into fresh fowls[72]. The fowls did not appear to be affected, and it looked as though the cultures were useless. Pasteur intended to throw them away and start again, when it occurred to him to inoculate the same unaffected fowls with a fresh and deadly culture of the bacilli responsible for the disease. The fresh culture proved deadly enough with ordinary fowls, but most of those which had previously been inoculated from the old culture with-stood the disease. During the accidental interruption of Pasteur's work, something had been happening in the ageing cultures. The capacity to produce disease was be-coming attenuated, but the capacity to give immunity was retained. Pasteur proceeded to extend his great discovery to the prevention of anthrax in sheep and cattle, with a success that astonished the world.

The combination of a good scientist and good luck brings good results. It often happens that a single scientist has very good and very bad luck, and by his success and failure shows the large element that chance must always play in scientific investigation. Mendel provides a perfect example. He chose peas for his experiments on inheritance. The choice was very fortunate, for the results were clear-cut, and he was able to lay the foundations of the branch of biology that bears his name. He also studied inheritance in hawkweeds (*Hieracium*), and devoted a lot of time to it; but he could scarcely have had worse luck than to choose this genus, and to be urged on by Nägeli to persevere with it[80a]. Nearly twenty years after his death it was discovered that in the genus *Hieracium* the seeds are often formed without fertilization, and the progeny therefore resemble the seed-parent without any influence from the pollen of another plant. A more unsuitable genus for the study of biparental inheritance can therefore hardly be conceived, though Mendel did not and could not know the reason why. He hoped that the laws he formulated as a result of his study of the pea were universally valid, and so indeed they are in ordinary biparental species of both plants and animals. Mendel naturally thought his hawkweeds biparental. He was wrong, and it was very bad luck, since uniparental reproduction is rare. If Mendel had worked only with hawkweeds, he would have been a potential but not a recognized genius.

Discoveries that benefit mankind in a practical way are produced liberally by a free science, expanding always in whatever direction expansion is possible. Constrain science within the strait-jacket of close attention to material needs, and it cannot breathe freely enough to serve mankind either culturally or practically. H. G. Wells has realized the futility of research planned *ad hoc* towards stated ends. He tells how under certain circumstances he might have become a very wealthy man. He deplores the possibility that he might even have "endowed and dominated futile *ad hoc* research"[141a]. It is encouraging to know that such an influential man realizes that planned research is futile. The fact should be more widely known. If one's friend has an inoperable cancer, he is treated with radium or X-rays[1]. It is relevant to remember that radium was discovered by a man and his wife who were interested in radioactivity, and that knowledge of X-rays grew from the studies of those who concerned themselves with the phenomena of electrical discharge *in vacuo*. It may be anticipated that when another great discovery comes to bring help to sufferers from this disease it is likely to come from an equally unexpected quarter. Those who wish to make it come quickly cannot know how to act, but the best suggestion one can offer is the endowment of free research in various sciences. Cancer can only be completely conquered when the basic sciences have progressed far enough for full understanding. To pour out money on clinical medicine while leaving free research in the underlying sciences less well endowed is a short-sighted policy.

The active investigator should not be required continually to justify his work before others. Most good research workers try a certain number of long shots—rather crazy-sounding experiments which they do not like to admit even to their friends. Darwin, who tried them freely, called them "fool's experiments"; Bateson, "fanciful follies". Nothing exemplifies better than these the lack of understanding of the minds of research workers shown by those who demand rigid planning in advance. Bateson notices a kitchen mincing-machine for the first time. He is at once struck with the possibility of turning the instrument to scientific use. He orders conifer seedlings from a nurseryman, and immediately a fanciful folly is in full swing: the handle is rotated and the plants passed through to surrender their resins[14].

76 THE SCIENTIFIC LIFE

This was the man whom Haldane has called "the greatest geneticist whom England has so far produced"[58]. He did not sit down and draw out elaborate plans to look imposing before a committee of planners: he tried his fanciful follies when he wanted to do so. Sometimes such experiments succeed, sometimes they fail. It is a matter of luck. If they fail and teach nothing, the research worker should not be required to account for their failure to anyone else.

It is sometimes argued that if chance has played such a large part in discovery in the past, we should do well now to exclude it by planning for every possible contingency, so that ordered advance may take the place of unexpected discovery. The answer is that chance alone never makes discoveries. It is the combination of chance with the prepared mind. Chance we cannot plan for: the prepared mind we can select. If we choose for our research workers only those who will follow some one else's plan with docility, science will stagnate. The facts cannot be put better than in the words of Dr. J. I. O. Masson, Vice-Chancellor of the University of Sheffield. "In time of peace to insist, as various people have tried to do of late years, that the active curiosities of original minds shall be turned aside to topics selected by non-practitioners or lay-arbiters as bearing on current public problems; to constrain into 'practical' themes men of highly specific abilities, whose reason for living hard and (by newspaper standards) inconspicuous lives is that they crave to know certain sorts of unknown things and can endure the long process of finding out: all this is a kind of regimentation which would frustrate its own ends. For the mind of the genuine researcher (as distinct from imitations, here negligible) simply will not work productively upon things in which it is not technically interested, or on 'problems of the day' which he knows are too crowded with shifting and arbitrary variables to be resolved by means of his specific aptitudes. Each researcher is a key that opens only the range of locks which it fits. His own results are normally what the public calls 'useless', because they are beyond the horizon of even the educated contemporary layman, and because the integrated effects of them and their like will not reach the surface of public apprehension for a long time to come"[93].

Let us have one plan only: the plan to choose as our investigators those active, independent, untamed people who, by a combination of alert and prepared minds, intense enthusiasm, luck, fools' experiments, and more ordered schemes of their own, are capable of making real discoveries. Such people will be ready for the arrival of an idea at any time of day or night. Naturally they will devise a method of research. They will be better equipped than any scientific dictator or planning committee to see how their own particular aptitudes can be put to the task of expanding the bounds of knowledge. Working along a line devised by themselves, they will be ready and free instantaneously to switch their energies in a new direction, if the unexpected appears. Good men are what science will always thrive on, not dry, dictated plans.

Chapter 6

The Origin and Nature of Science

"The greatest truths are perhaps those which being simple in themselves illuminate a large and complex body of knowledge."
ALEXANDER[3]

The University of Chicago. A time and motion comparison on four methods of dishwashing : a dissertation submitted to the graduate faculty in candidacy for the degree of master of arts. [80]

Curiosity

So far as their sense-organs will allow, animals apprehend truly those attributes of objects and phenomena of which they take cognizance. Truth has selection-value. An animal could not survive which was personally insensitive to truth, and went in one direction when the evidence of its senses suggested that it would find its food in another. This perception of truth is not limited to what immediately concerns it: an animal is aware of objects in its environment which have neither positive nor negative value for it. When, in the evolution of human beings, the power of speech began to be developed, primitive man maintained his animal ancestor's apprehension of truth. He was unpractised in verbal deception, and it is only reasonable to suppose that when he conveyed information to another person by the first rude attempts at speech, he conveyed, so far as he could, the truth. Thus was made possible, many thousands of years later, the birth of science. If the sense-organs of animals and early man had evolved in such a way that only useful and harmful objects could be apprehended, science could never have been born.

To say that truth has selection value is of course not the same thing as postulating that any thought, originating from sense-perceptions of the environment and having selection value, is necessarily true. As Mr. R. Brown has pointed out to me, the beliefs of various Eastern peoples about pork have selection value, though they are false. Selection value provides no criterion of truth; but no such thing as appreciation of truth could have arisen if truth did not possess selection value far more usually than falsity.

Evidence is beginning to be collected showing that the lower animals recognize objects by tests which are different from those we apply to the same objects, but there is no reason to suppose that our remote half-human ancestors in Pliocene times differed materially from ourselves in this respect, for the sense-organs and the parts of the sub-human brain concerned with perception must have been fundamentally the same. Early man, it must be inferred, saw roughly as we see: and within him was the urge to see more widely. This point has perhaps been insufficiently stressed. Before he could get about otherwise than on his own legs he had already spread over nearly the whole of the habitable surface of the globe, though he could not fly like a bird or even gallop like a horse. This world-wide distribution of a single species of a heavy terrestrial animal is unique, and the mere necessity to look for food and shelter is unlikely to have been a sufficient stimulus to have caused early man to walk voluntarily into the barren and inhospitable areas which he occupied. He was inquisitive about the environment as a whole, as many animals are. One has only to watch a cat investigating an unfamiliar object in her haunts to see an embryonic rudiment of

a curiosity resembling man's. J. G. Crowther[26] says that
curiosity is a sublimation of the desire for power, but
somehow it is strange to think of fishes indulging in sublimation, yet fishes show curiosity[117]. The Primates go
infinitely farther than fish or cat. Darwin saw a monkey
investigating a lens and succeeding in placing it at the
proper distance from its eye to obtain a focus[37]. Everyone knows how inquisitive monkeys are; the fact is inescapable, and no light is thrown on the subject by saying
that they are only sublimating power-urges. We live today
because our ancestors were inquisitive and went—because
they wanted to go—outside their haunts to find unfamiliar
objects and investigate them. St. Augustine condemned the
"vain and curious desire of investigation", but he would
not have lived if his ancestors had not possessed it.

The investigation of objects carries with it the necessity
to generalize about them, and it is inescapable that the
capacity to generalize must have had a selection-value, for
an individual which regarded each object as *sui generis*
would be at a disadvantage in the struggle for existence
against others who could infer the properties of objects
from their resemblance to those already known. The necessity to generalize may have appeared subjectively to early
man, as it does to us, in the form of pleasure in finding
order. Pleasure in curiosity and in the satisfaction of curiosity
combined with pleasure in generalization were the rudiments
from which has grown that part of science which may be
called "natural history" (see p. 92). Curiosity by itself is
of course not a mainspring of science. Completely un-
scientific people are often very curious about all sorts of
matters which cannot be welded by generalizations into an
ordered system of demonstrable knowledge.

Every instinctive feeling varies in its intensity in different
people. In conformity with what is known of inheritance,

one must suppose that the variations in intensity are governed partly by inborn factors and partly by the environmental conditions, especially of youth. It would be strange indeed if everyone had identically the same desire to enquire and to generalize about the results of enquiry, and in fact one finds the whole gamut of variation in present-day life from the man whose pre-occupation is the football-pool to the greatest minds of modern science. No one can tell whether heredity or early environment generally plays the larger part in the production of a scientist: however favourable the inheritance, extreme unsuitability of early environment will crush it unless one has the innate ability of a Faraday or a Mendel, while if the innate desire to find out and generalize is conspicuously small, no amount of encouragement or provision of lavish facilities will avail.

Although curiosity and the desire to generalize are the mainsprings of the scientific mind, yet clearly they cannot alone create a scientist: originality and intelligence and perseverance are necessities, and the finding of pleasure in the use of the hands in delicate manipulations is very nearly necessary. The presence of this combination of qualities in a higher degree of development necessarily differentiates the natural scientific investigator from those in whom they are relatively deficient.

Observation and experiment

The demonstrable facts, whose observation, recording and grouping constitute the province of science, present themselves to the observer in two different ways: they may be either explicit or implicit. My choice of the word explicit does not indicate that the study of this group of facts is easier than that of the other, but only that they are ready for observation without the necessity for the observer to produce artificially the conditions under which they will

F

arise. The explicit facts require observation only, the implicit ones require experiment. Some explicit facts, such as the minutest striations on the shells of the microscopic plants called Diatoms, are very difficult to observe; some implicit ones, like the regeneration of a flatworm when a part is cut off, are very easy.

The only difference between explicit and implicit facts lies in the circumstance of whether the observer interferes with the subject of his study in such a way as deliberately to change it artificially; it depends on whether the human will has operated to change the object observed. When the observer mounts Diatoms in special media to make their striations more apparent and uses a microscope to view them, he is only taking steps to observe explicit facts. The use of reagents and instruments does not make his method experimental. If, however, he were to crush the Diatom to study the nature of the fracture of the shell, he would be studying implicit facts or making an experiment. It is not easy to understand but it is nevertheless true that some scientists are more attracted by the study of explicit and others by that of implicit facts, though there is no essential difference between the two kinds. It is in biology that the two kinds of facts give the strongest impression of being really different. The organism is either studied as it appears in an ordinary environment, or else in an environment which produces changes in structure or function which would not have originated apart from the operation of man's will. These changes are fascinating to some minds, while to others they are much less interesting than the explicit facts of the structure and behaviour of the organism in its "natural" state.

There is no intrinsic value which would place observational above experimental methods or *vice versa*. Some parts of science require one method and some another: many require both. The notion that science is served only by experiment is fallacious. Waddington[139] says that "science is concerned to discover how things work, and its test of truth is that it can make them work as it wants to". Thus the whole heritage and future of observational science—astronomy, for instance, and palaeontology—is discarded, and all the emphasis is placed on experiment and human "want".

The experimental method is, of course, an essential part of many branches of science. Modern botanists would be horrified to read an old-fashioned statement that botany is an observational science. The great American biologist Morgan has well suggested[104] that it would be as absurd for biologists to rely wholly on observation as for physicists and chemists to confine their activities to the study of such natural phenomena as thunderstorms and volcanic eruptions. This must be readily allowed, yet such a thing as observational science is necessary, and at the end of this chapter I shall suggest that it may be helpful to classify the largely observational sciences together as "natural history". I once knew a man who used to say, in answer to non-scientific strangers who asked what his job was, that he was an experimental biologist. He enjoyed their bewilderment. I confess that I was as bewildered as the strangers, for a man might as well deliberately make scientific research more difficult for himself by cutting off one of his hands as discard purely observational methods when they will help as guides to understanding. In both observational and experimental science it is the answer to a question that is sought, not the fulfilment of an attempt to make things work as one "wants". The introduction of the word "want" into a definition of science brings to mind a question, indicative of a naïve misapprehension of what science means, which scientists may sometimes find themselves asked to answer: "What are you trying to prove?"

The definition of science

How, then, may science truly be defined? The aim of natural science, as Findlay[47] has said, "is to acquire as complete a knowledge as possible of the material universe; of the objects, materials and phenomena and the relations between the phenomena which make themselves known to us or which we apprehend by means of our senses". By following this finely-stated aim a structure of knowledge is built up which, as A. V. Hill[66] remarks, "is approved by all sane men"; though it would be wise to add that only the sane men who have studied it are in a position to give their approval. Approval is actually an understatement of the feelings of men towards that "body of valid ideas"[110] which constitute science. Polanyi, in his great though short statement of the true significance of science, has remarked how nothing has proved so permanent as the systems of science. "Waves of civilization have come and gone over Mesopotamia, Egypt, and Europe, and while their creeds and laws, and often even their crafts, may have been forgotten, their contributions to systematic science have been preserved. It seems that an ordered framework of ideas in which each single part is borne out by the cohesion of the whole is of supreme attraction to the human mind. Struggling for a foothold in a shifting world, the mind clings persistently to these rare structures of sound and consistent ideas. It is in these structures, accordingly, that all scientific interest resides"[110].

Whether we speak austerely, with Hill, of "approval", or enthusiastically, with Polanyi, of "supreme attraction", we mean the same thing: science has a human value as an end in itself.

Science, as Alexander has said, is not "a mere repetition of facts; that would be chronicle and not science"[3]. A discovery is "important" in science when it illuminates a large field of knowledge: the larger the field it illuminates, the more important it is. Often the great discovery is attractively simple. "The birth of the greatest truths is for the most part an event in which some simple statement of the widest generality emerges out of a great mass of more or less separate pieces of knowledge to which it suddenly supplies a clue or enables it to be taken in at once in a flash"[3].

J. G. Crowther[26] has defined science as "the system of behaviour by which man acquires mastery of the environment". We should laugh if a man set out to define "army" and gave a good definition of "navy" by mistake; yet this is exactly comparable to what Crowther has done. His defininition is a good one, but not of science: it is a definition of technology. It is important that no energy should be wasted on an argument about words, and it is therefore convenient to say that one of the contrasted subjects is called science or pure science or fundamental science, and the other is called technology or applied science or (by those who distinguish between the two latter) technology and applied science. I do not mind at all which of these words or phrases is used, but in this book I generally use simply "science" and "technology". Technology serves man's material needs directly, and his spiritual needs indirectly (by providing, e.g., the ink and paper used by writers and composers). The uses of science are discussed in the next chapter. Here it must suffice to say that its principal uses are two: to serve as an end in itself, like music, and to form a basis for technology.

Just as the concepts "boy" and "man" are useful and generally valid, so are the concepts science and technology, despite the fact that the distinction cannot always be made with precision. There are, however, various guides on which some reliance may be placed in doubtful cases.

Science is essentially inductive. Facts are collected so that generalizations may be drawn from them. Technology, on the contrary, may almost be called deductive: use is made of the generalizations of science to deduce how material human wants may be satisfied. Those working deliberately towards material human welfare often make discoveries which are of scientific value: in so far as they do that, their work is not technological. Pharmacological research provides an example. There is much that is not scientific about our knowledge of the action of drugs, but when the facts compel a generalization, the subject is as scientific as any other. For instance, the knowledge that certain drugs stimulate the endings of the sympathetic nervous system and others those of the parasympathetic system is a part of science. It remains valid and interesting whether such drugs are used in pharmacy a hundred years hence or not.

That brings us directly to a second test of distinction. In science, every discovery is of permanent value, and serves as a basis for other discoveries. In technology discoveries are often of ephemeral value. When gramophones are invented, the musical box becomes less interesting. No fact discovered by technologists can be untrue, but its interest can wax and wane exceedingly in response to human needs and even whims. The test of the importance of a technological discovery is service to potentially-ephemeral human needs; the test of importance of a scientific discovery, as we have seen, is its capacity to link up and illuminate other facts, and that can never change.

Technology is deliberately eclectic, while science aims always towards completeness. If science disappeared and only technology remained, knowledge would be a hotchpotch instead of an ordered system. Special investigations would continually have to be made to find out basic facts which would have been common property long before, if science had survived.

The indulgent reader may allow a little parable which

serves to distinguish science from technology in a way that will in some cases be found helpful. Suppose that a terrible disease ravaged mankind and human beings became extinct. Suppose that the brain of another animal evolved until it equalled that of man. Imagine that this happened, for instance, to the hippopotamus. (If the reader is inclined to smile, let him ask himself whether a visitor from Mars would find the grant of intelligence to a hippopotamus or to an ape the funnier.) Eventually the hippopotami would build up a science, even though difficulties in manipulation of apparatus would make progress rather slow. Atoms and molecules would be discovered; the structure of the benzene ring, even electrons and quanta would at last be disclosed. Meanwhile the hippopotami would be constructing special mechanical carriages for their more convenient and rapid transport from place to place, and discovering drugs to alleviate the ills to which their flesh was heir. Let us now look at the sum-total of their knowledge. Their engineering and pharmacy would be quite different from ours: their chemistry and physics, on the contrary, the same. Chemistry and physics are sciences, engineering and pharmacy are technology.

It would take us too far to pursue these imaginary events in detail, but this may be noted. If the hippopotami discovered generalizations about the actions of drugs on the sympathetic and parasympathetic nerve-endings, these, as we have seen, would be science; but their study of social hippopotamology could only be regarded as unquestionably scientific if it were related to knowledge of the social life of other animals.

Technology, of course, itself brings advantages to science. I never cease to be genuinely thankful to the engineers whose work has made light, heat and water available in my laboratory at the touch of a switch or tap. These conveniences are as useful to me as typewriters and printing-presses to writers. They are invaluable, although they provide no inspiration.

Nothing in this book is intended as an attempt to minimize the magnificent achievements of technology. Such a stupid attempt would in any event be futile. I have myself worked for years in both technological and scientific research. I know the radical differences of atmosphere and approach in the two subjects and the special attractions and difficulties of each. To myself science makes the greater appeal. This is a personal, subjective matter. Edison and Marconi may be cited as very great technologists, Faraday and Darwin as very great scientists. It is wise to refrain from arguments as to which is the higher form of activity. The energy spent on such an arid discussion would be better used in seeking to find the right places in research for the right people, according to whether their personal inclinations are towards science or technology.

Technology should not be magnified at the expense of science. In recent years the British Association has shown a definite tendency in this direction, and has laid itself open to the charge of having become an Association for the Advancement of Technology. Such an association might be an admirable one, but it should not arise at the expense of science. Anyone who magnifies technology at the expense of science is like a man who cuts down a tree because he is so anxious to get the fruit that he forgets the necessity for roots and stem and the potential beauty of next year's flowers. Science is a perennial; the roots and stem must be allowed to survive, and the ground be cultivated, so that flowers and fruit may come again another year.

Logic and mathematics cannot, I think, be included as component parts of science, despite the important contributions they make to certain aspects of scientific work. It is a striking fact that persons untrained even in the elements of logic are usually logical enough for all the purposes of ordinary scientific research, though it would be far from my intention to belittle the subject, which Herbert Spencer regarded as forming, with mathematics, the abstract division of science[122]. It seems desirable, however, to separate mathematics and logic more sharply than this from that study— science in the narrower sense—which is concerned with information received through the sense-organs. If a person blind from birth and lacking the sense of touch were sufficiently intelligent and could picture to himself such imaginable things as separate objects and such unimaginable ones as points and lines, he could dictate the whole of mathematics (except perhaps probability) out of his head without any instruction or the necessity to know anything about the universe. Logic and mathematics, as Driesch[40] says, constitute the "realm of mere meanings", which he separates from "the realm of Nature", which is equivalent to science. The mathematician, the same author remarks, investigates *the* sphere"; the scientist investigates "this *one definite sphere"* (which, it might be commented, is never spherical).

The argument as to whether mathematics is a part of science is, of course, largely an argument about words. The use of sensory perception in the one study, and not in the other, is probably sufficient justification for the adoption of a single word to cover all branches of the former study, even though the process of generalization is shared with the other and has indeed been carried much further in it. The only single word which has been used to mean the recording of sensory perceptions of the universe and generalization about them is *science*, and it may be suggested that the wider meaning sometimes attached to this word has on the whole more disadvantages than advantages.

A classification of the sciences

Any classification of the sciences must be arbitrary and convenient rather than demonstrably true, but I want tentatively to suggest that it may be helpful to classify them in four groups: basic, intermediate, expanded and eclectic. The basic sciences are physics (including mechanics), chemistry, astronomy, geology and biology. Each of these is of course divided into basic sub-sciences: thus the study of geographical distribution is a sub-science of biology. Biochemistry is an obvious example of an intermediate science. It is difficult to decide whether such sciences are genuinely intermediate, or whether they have only obtained their intermediate position because science has chanced to develop in a particular way. The fertilizing contact of people trained in different schools of thought has probably been responsible for the rapid growth and extraordinary interest of intermediate sciences, when at last people begin to bridge the gulfs. Anthropology and psychology are expanded sciences: that is, they constitute parts of the basic science of biology, but they have necessarily been so greatly expanded by the peculiar status of man and mind in the universe as to deserve separate recognition. Pharmacology, pathology, physical geography and meteorology may be regarded as eclectic sciences, not simply intermediate between certain of the basis sciences, but drawing upon chosen parts of two or more in such a way as to illuminate special fields of investigation.

Pharmacology is a very special kind of science. It could be set apart as technological if the inductive method had not been used in it with such success. It is very exceptional in the sharp limitation of its field. A worker in any other science cannot fail to be surprised when he reads about drugs of plant origin in book after book of pharmacology and notices no mention of the significance of those substances in the lives of the plants themselves. Such a deliberate limitation of interest is as surprising as the fact that scarcely a single valuable plant drug was first tried on the recommendation of a pharmacologist: nearly all were first used by simple, uneducated people. The inquisitiveness and initiative of the nameless discoverers are truly remarkable and would scarcely be thinkable in present-day Europe, where discovery is unfortunately left by common consent almost entirely to professional laboratory workers.

Rather a sharp cleavage between two kinds of outlook divides the biologist from the non-biological scientist. The cleavage is caused by the appearance of purposiveness in living organisms. An example will illustrate the point. Haemoglobin, the red colouring matter of the blood of man and many animals, is readily oxygenated and the oxygenated compound parts readily with oxygen. This fact is regarded by biologists as very important, because it is made use of by so many animals for the transport of oxygen to the tissues. Haemoglobin has, however, another very curious property. It can act as what is called a "peroxidase": that is, it can decompose hydrogen peroxide with the production of active oxygen. This is something which apparently never happens in the blood of living organisms. Biologists therefore regard this property of haemoglobin as much less interesting than the other, although both equally are properties of a substance which is only formed in living organisms and which therefore belongs to their realm of investigation. Chemists and physicists look on facts and phenomena in quite a different way. They are just as interested in reactions which never occur in natural circumstancs as in those which are occurring all the time (e.g., in volcanoes). The biologist is interested in the organism as a going concern. He regards those attributes of its component parts and substances which do not affect its survival and reproduction as far less important than those that do.

Natural history

There exist criteria, rather convenient for some purposes, by which the whole of science may be cut right across and separated into only two divisions. Of these one may be called general analytical science and the other natural history. The latter is a valuable phrase, though it has been used in different senses. In common speech it is sometimes used rather loosely to mean those parts of biology which can be pursued without special training or the use of special instruments. Karl Pearson[108] has used it in an extraordinary way, which I shall not adopt: for him, natural history is not a part of biology or vice versa, but is contrasted with it as a separate part of organic science. This use of the phrase presents no advantage and has no historical basis. Thomas Hobbes[67] defined natural history as "the history of such facts or effects of nature as have no dependence on man's will; such as are the histories of metals, plants, animals, regions, and the like". He contrasted this with civil history, the history of the voluntary actions of man. Everything in modern science except psychology and social anthropology would be included in his definition of natural history. The great French zoologist, Baron Cuvier, narrowed the definition considerably. He divided all science into "physique générale" and "physique particulière"[27]. The former comprised what we now call physics and chemistry, and Cuvier defined this part of science as that in which bodies may be isolated and reduced to their utmost simplicity, and their properties brought separately into action. By particular physics he meant the study of those objects which do not allow of rigorous calculation or of precise measurement in all their parts. The phenomena here occur under circumstances which do not permit an experimenter to reduce the problem to its elements: it must be taken whole, as, for example, when one examines the activities of a whole living organism.

Cuvier used the words "l'histoire naturelle" as synonymous with particular physics, and stressed the relative importance of observation in this sphere in comparison with experiment. He thought it best to range astronomy and meteorology with general physics. For him, then, natural history may be taken to mean botany, zoology, and geology; but he would presumably not have included present-day physiology, in which analytical methods and experiment play so large a part.

There is an element in Hobbes's definition which is helpful to the concept of natural history. Man's deliberate will is such an extraordinary phenomenon in the universe that attention may properly be paid to it in a classification of the sciences. Though probably it cannot be separated in kind from the mental activities of the higher animals, yet it transcends them so much in degree that one may readily allow a special adjective, "natural", for objects and phenomena which are not at all or not greatly influenced by it. T. H. Huxley[77] said, "Nature is, of course, the totality of all laws", and I remember reading somewhere that the natural is simply that which occurs; but if we make the word co-extensive with everything, we extend it to the point of depriving it of meaning.

The phrase natural history, then, may be used to cover the investigation of particular explicit objects and phenomena, which can be studied as wholes mainly by observation without much or any human interference; e.g., a thunderstorm, or the movement of the moon, or the normal growth or habits of an animal. General analytical science, on the contrary, is concerned with general rather than particular, and implicit rather than explicit facts; it usually requires human interference by means of experiment. Examples are the study of the chemical properties of an element or of the behaviour of gases under various pressures and temperatures.

In this broad grouping of the sciences, we may classify under the heading of natural history a large part of biology together with geology, meteorology and astronomy. The latter two subjects are so much concerned with particular rainfalls and particular stars that I cannot follow Cuvier in excluding them from natural history. (Cuvier himself felt doubts about these two subjects, especially astronomy.) Physics and chemistry are the general analytical sciences *par excellence*. The extent of the use of experiment and analysis in genetics and physiology bring these biological subjects within the group of analytical sciences. The tendency in animal physiology is to disregard the activities of the organism as a whole, and to study it without careful consideration of its relationship to other animals, its evolution, its geographical distribution or its natural habitat and habits.

It would be far from my intention to suggest that experiment is not used in biology except in genetics and physiology, or that observation without experiment is not used in physics; but I think that most scientists may be grouped according to whether their main inclinations are to general analytical science or to natural history. It is rather because I believe that there are two main kinds of scientists, than because there are necessarily two main kinds of science, that the division of science in two is suggested. Those who may be called the natural historians derive their impetus to research, I believe, directly from their ancestors' interest in the environment, with which subject the first part of this chapter is concerned. The interest exists in everybody;

when it is very strong, the person tends to become a research worker in one of the natural history sciences or an explorer in the geographical sense. The analytically-minded person, who finds little interest in wholes and tends to be interested in substance rather than form, in the general rather than the particular, in the implicit rather than the explicit, is more difficult to account for on an evolutionary hypothesis. While we see the exhibition of curiosity about the natural environment in a wide range of animals, deliberate interference with nature for the purpose of gaining knowledge is almost confined to man, and probably evolved as the capacity to use tools was being perfected. Elements of this kind of mental make-up exist in everybody, but those in whom it is particularly strongly developed tend to become research workers in physics or chemistry. Although interest in the environment and in its alteration by tools were both useful to primitive man, yet their direct usefulness brought man as short a distance on the way to scientific culture as the love-cries of the ancestral ape have brought him in music (though it must be admitted that not everyone can sing an octave in semitones as accurately as certain anthropoid apes[35]). There is no evidence that the potential brain-power of man has evolved during the historical period, nor any knowledge which would suggest that such evolution could have occurred. By the use of tradition, first oral and then by writing and finally by printing, inquisitive and analytical minds have passed on, from generation to generation, the fruits of their studies, and thus knowledge has been built up to the magnificent heritage of present-day science. The stimuli have been the primitive urges which were useful to prehistoric man, but scientists have used those urges differently from their remote ancestors. Just as speech no doubt evolved on account of its practical usefulness, but has given us our heritage of literature, so science has transformed itself in such a way that the greatest discoveries have not been made in response to material need. If each generation of man had always to start where the last generation started, the primitive urges would get him no further than the fulfilment of immediate material needs and increase in the range of his species. It is through step-like progress, made possible by each generation possessing at the start the treasures of the one before, that man has been able to transcend the functions of his primitive urges and produce modern culture.

Neither general analytical science nor natural history is superior to the other. Scientists should be ready, as we have seen, to use observation or experiment or both according to the nature of the problem in hand, and not to tie themselves down as to the particular method of approach. There is some tendency for general analytical science to be given a higher standing than natural history. Indeed, as my friend Mr. Charles Elton remarked to me long ago, one can arrange some of the sciences in an order (physics, chemistry, biology, anthropology, psychology) in which the exponents of each tend to regard themselves as superior to those of the next. More remarkable still is the fact that scientists in all groups except physics (which has no "superior") tend to look with a certain amount of awe on the sciences before them on the list. These ideas are crystallized in the statement, attributed to Rutherford, that science consists of physics and stamp-collecting. This is an epigram intended to mean that particular objects are uninteresting: it is the extreme view-point of a general analytical scientist. It is true that chemistry rests partly on physics, biology on chemistry, and anthropology on biology, and that psychology must eventually rest on anthropology and biology; but this in no way invalidates demonstrable facts in any science. A good classification of a group of animals is valid although we cannot yet describe the chemical differences in the genes which make one animal different from another. Psychology has produced valid concepts which cannot yet be linked up with brain physiology. True science never arbitrarily limits itself, and psychologists and brain physiologists will no doubt work gradually towards each other's spheres; but it is neither necessary nor desirable to wait until a gulf is bridged, before cultivating the land on the other side. Science would be very poor in total content if we dismissed everything that cannot be stated in terms of electrons, protons and quanta. Feelings of superiority and inferiority do nothing for the welfare of science and should be forgotten. Each research worker should take his place in that sphere in which he feels that he can make the greatest contributions to knowledge.

Chapter 7

The Uses of Science

". . . until a sufficient foundation of pure science has been successfully laid there can be no applied science. Real progress comes from the pursuit of knowledge for its own sake."
TILDEN[132]

SCIENCE, as was pointed out in the last chapter, has two principal functions. It exists as an end in itself, finding its natural place in culture alongside music, art, literature and philosophy. It exists also as the only possible base from which technology can advance. These two functions are both useful.

There has been much misunderstanding about the word "useful". People have often restricted the sense of the word to make it synonymous with "beneficial in a material way", and some have even further confused the subject by using the word "utilitarian" with the same meaning. Since utilitarianism is intimately connected with the philosophy of J. S. Mill, it greatly simplifies argument if the words are used as he used them.

Mill[99] made it clear that whatever increases pleasure or decreases pain in any sentient creature is useful or utilitarian. He maintained that usefulness has no other meaning. He found it impossible to oppose the agreeable or ornamental to the useful. He allowed, naturally, that some kinds of pleasure are more desirable than others. There is a sense of dignity possessed to some extent by all human beings, which leads them to prefer one kind of pleasure to another.

It is particularly the kind of pleasure which man does not share with animals that men regard as the highest in quality, so that they consider it "better to be a human being dissatisfied than a pig satisfied". As Mill said, no intelligent person would willingly consent to be a fool, even if he were fully persuaded that the fool is better satisfied with his lot than he is himself. People who have experienced pleasures which employ their higher faculties give a marked preference to these over the lower pleasures, such as that of eating.

When anyone is undernourished or lacks sleep or shelter or warmth or health, nothing is more useful to that person than the provision of material help; but when everything has been rectified, there remains a void unless something can exist for that person as an end in itself. In the modern world people fight shy of the concept of the end-in-itself with a scarcely comprehensible evasiveness. In art they will even pretend that beauty is solely functional, though everyone who has a garden and grows flowers without making material use of them (e.g., for pharmacy) is a silent witness that the appreciation of beauty is in fact an end in itself; for not one gardener in a hundred is interested in the function of flowers, which is the achievement of cross-fertilization. Those who say that nothing except suitability for material ends gives beauty will also find themselves in difficulty over music, for apart from its outstanding performance at the siege of Jericho, it has never been of material service to man except as a means of keeping soldiers in step and stimulating their martial ardour.

Those who appreciate science as an end in itself are subjectively aware of its value to themselves: it is to them what music is to the musical. In Chapter 9 a plea is made for the wider spread of scientific culture, so that the greatest possible number of people may benefit from it as an end in itself. This aspect of science has not been sufficiently recognized.

Scientists would be selfish people if they wished to keep the culture of science within closed walls: indeed, if they succeeded, the subject would scarcely deserve to rank as culture. Not every scientist could or should be himself a disseminator of scientific culture, but in Chapter 9 ways will be suggested in which that culture could be spread so as to give the maximum benefit to humanity.

Galen, nearly eighteen centuries ago, recognized the importance of knowledge of anatomy not only for the physician but also, as he said, for the philosopher[129]. That was his way of acknowledging its value as a final end. This acknowledgment springs from subjective and not objective experience and cannot therefore become a public truth. If a man says, "I appreciate the music of Beethoven apart from the possibility of its use for keeping soldiers in step", we cannot by any objective means prove him truthful or a liar. It is the same with science: no satisfying proof is possible. The fact remains that the individual scientist knows certainly what a tremendous value science has for himself as a final end, and he infers from the disinterested enthusiasm of his colleagues that it appears in the same sort of light to them.

One knows that scientists are absorbed in science as composers and artists are in music and art. J. A. Crowther (who must not be confused with J. G. Crowther) has written[25], "In spite of the difference in their working media, the man of science, philosopher, poet, and artist belong to the same category; as anyone who has had the privilege of working under any of the great men of science of the age can easily realize." Even when Perkin had discovered aniline dyes (see p. 64) and was exploiting them commercially, his heart remained devoted to science for science's sake[133]. He always continued research in science apart from his technological work, and in the end reverted to it entirely. That is the usual attitude among those scientists who make great contributions to material human welfare. In his speech at the opening of the Faculté des Sciences at Lille, Pasteur told his audience that it would not be their part to "share the opinion of those narrow minds which scorn everything in the sciences which has not an immediate application"[136]. Nearly ninety years have gone by since Pasteur spoke those words, but narrow minds are still with us, and indeed they are much more influential in our time than they were in his.

The first of the five propositions to which the members of the Society for Freedom in Science adhere is this: "The increase of knowledge by scientific research of all kinds and the maintenance and spread of scientific culture have an independent and primary human value." This sentence expresses so exactly the feelings which are known to animate most scientists, that it would be unnecessary to labour the point if certain thinkers had not adopted rather an equivocal position which demands careful consideration. Although their thesis contains valuable elements, yet it can serve to confuse the issue.

Bertrand Russell[118] is the chief expositor of a function for science intermediate between its service as an end in itself and direct application to material ends. It would appear that he does not himself accept the function of science as a final end. "Perhaps the most important advantage of 'useless' knowledge", he says, "is that it promotes a contemplative habit of mind." Immediately one asks, "So that one can contemplate what?" If science is the answer, then science is allowed to be an end in itself. But apparently this is a short-circuit not intended by Russell, for he says that mental cultivation "gives other interests than the ill-treatment of neighbours, and other sources of self-respect than the assertion of domination". This is very true and valuable, but it seems that he is not regarding science as an end in itself, but only as a means of preventing quarrels. He also regards science as a means of preventing unhappiness in general, which it certainly can be, though this is a curiously negative and partial exposition of the possibilities of scientific culture. "A life confined to what is personal is likely, sooner or later, to become unbearably painful; it is only by windows into a larger and less fretful cosmos that the more tragic parts of life become endurable." These thoughtful and helpful words actually tend to obscure the positive concept of science as an end in itself. Raymond Mortimer has written in much the same vein about the appreciation of nature, though he was not thinking specifically of its scientific appreciation[105]. Mill himself came near to the same position when he wrote, "Next to selfishness, the principal cause which makes life unsatisfactory is want of mental cultivation"[99]. Hopkins stresses a positive value in science which may be associated with Russell's rather negative but valuable one. He points out[73] the debt that is owed to Darwin, Lyell and T. H. Huxley for their work in spreading a belief in the value of intellectual honesty.

It may be concluded that art, music and science have a first function as ends in themselves and a second one in making the mind less petty and more honest, and thus helping in problems not connected with the subjects themselves, e.g., by giving consolation in unhappiness. Beyond these two functions science has the third and very important one of forming a basis for technology. Art also has this function, in so far as it affects the design of materially useful objects, unless its effect upon design be regarded simply as an aspect of the first function. Music cannot claim to share the third function, since the only techniques which it affects are the internal ones of the manufacture of musical instruments and the acoustic design of auditoria. Science could not claim to share the third function, if it only provided a basis for the technique of manufacturing scientific instruments.

It would be futile to try to assess the values of the different functions of science. It must be admitted that nothing at all can be more stultifying to a discoverer than to have to think continually about whether what he discovers is going to be materially useful. Similarly there is absolutely no point in trying artificially to suggest material uses for studies whose intrinsic interest is their justification. For instance, there is an immense amount of knowledge of the extinct reptiles called Dinosaurs, and in fact it has been said that "to quote all the literature or to refer to all the Dinosaurs would require not a book but an encyclopaedia"[125]. The rise and fall of these astonishing animals is interesting to anyone. (Incidentally it may be mentioned that very good reconstructions of some of them are given in Disney's film, "Fantasia", in which they come to life in a convincing way.) Swinton's valuable book[125] on these animals is not helped but damaged by an almost pathetic attempt at the end to show that certain aspects of their study might perhaps be practically useful to the prospector or the mining engineer. This is comparable to the simile used before of keeping soldiers in step by the music of Beethoven.

It is rather curious that materially-minded people regard the first function of science as play, while a great scientist who laid the chief stress on that function used exactly the same word for the material use of knowledge. Archimedes regarded his extraordinarily ingenious mechanical inventions, for which he was famous, as merely "the diversions of geometry at play"[63], despite their real and highly appreciated material value. Knowledge apart from material uses was for him of enormously greater value. Some scientists approximate to one view-point and some to another, but no one can advance science by minimizing any of its values.

No one can say for certain that any piece of knowledge will never be of material use to man, but one can and must deny J. G. Crowther's statement[26] that "Nearly all scientific discoveries have proved of practical value". An enormous amount of discovery has been done, for instance, in connexion with the classification and geographical distribution of organisms, the migration of birds, mimicry, coral reefs, histochemical tests, experimental embryology and regeneration, that has never yet found practical application. These are only a few random instances taken from the zoological field. Some people are luckily so constituted that they will not agree to remain totally ignorant of such extraordinary phenomena as the migration of birds and the growth of coral reefs, simply because they are not convinced that the study of these problems will benefit man in a practical way.

Simple people, living close to nature, show an intense interest in natural objects and phenomena quite apart from their material use. I have found this both in the Sinhalese population of the vicinity of the Sinharaja rain-forest in South-West Ceylon and also among the savages of the New Hebrides in the Pacific Ocean. Indeed, a scientist of the kind which I have called the natural historian (see p. 92) would find himself more at home with these people than among the city-dwellers of his own country. I was once walking along a knife-edge ridge in the mountainous centre of the island of Espiritu Santo in the New Hebrides, when a conversation occurred which has stuck in my head ever since. I was two and a half days' arduous journey from the nearest white man, and my companions were the local savages and my native porters. The local people conveyed some information to me as we went southwards along the ridge. It had to be converted into pidgin-English by my porters and I shall here re-convert it into my own tongue. They were telling me that the rain which fell on my right hand would reach the sea on the west coast of the island, while that which fell on my left would follow an entirely different course and reach it in a huge bay on the north coast. Here, extremely remote from anything resembling what most people call civilization, I was receiving a physio-graphical lecture on watersheds! I had had an equally striking experience many years before, when on an expedition to the Banks Islands, to the north of the New Hebrides. My companion, Mr. T. T. Barnard, found that the natives understood the extraordinary habits of cuckoos, without having received information on the subject from white men. These examples serve to show the innate tendency of human beings to take interest in the environment, without regard to material usefulness. It is shown also by the extraordinary profusion of native names for living organisms, whether materially useful or not, in both the New Hebrides and Ceylon. In both places some of the natives are relatively uninterested in natural history, others intensely interested and in possession of a considerable amount of knowledge. Neither they nor academic botanists would agree with the remark of Socrates that he "had nothing to learn from the trees". Every tree has its native name, and its correct identification is a matter of interest, to be argued about seriously if necessary.

False indeed is the idea that things can only attract the study of human beings if they are of immediate material use. J. G. Crowther states dogmatically that "science is a product of human demands". "Science has evolved", he says, "from crafts and industry." Hogben[71] has written much in the same vein, and even gone so far as to reach a *reductio ad absurdum*. "From a landsman's point of view", he writes, "the earth remained at rest till it was discovered that pendulum clocks lose time if taken to a place nearer the equator. . . . After the invention of Huyghens the earth's axial motion was a socially necessary foundation for the colonial export of pendulum clocks." The suggestion that the only people (except sailors) who find interest in the fact that the earth revolves are those who can profit by the knowledge by making the necessary adjustments when exporting clocks towards the equator, is nothing short of fantastic. Hogben's words might have been written by an opponent who wished to satirize the opinion that science has no other function than to supply the material needs of man. The absence of a pendulum in a watch frees it from the effect of the earth's rotation, and Polanyi ironically and wittily suggests[110] that nowadays we may perhaps abandon the uninteresting idea that the earth rotates, because we mostly carry watches.

A proof that practical necessity is not required for scientific advance is provided by the many inventions of apparatus devised for purely scientific ends in the laboratory, and subsequently transferred to practical use in the outside world in ways not imagined by the inventors. As examples of such inventions Haldane[58] cites the telescope, cinematograph, barometer, gasometer, galvanometer and cream-separator. The microscope, as we have seen (p. 61), was perfected for the purposes of science and made available in its perfected form to the industrialist. We are here concerned not simply with the endless examples of practical applications of scientific knowledge to human affairs, but with the adaptation of actual instruments which were used by their inventors for the purpose of getting more knowledge about the universe. This exposes the fallacy of supposing, with J. G. Crowther, Hogben and others, that scientific results come only in response to material needs.

It would be reasonable to expect that one so insistent as Hogben on the directing influence of human need on science would exemplify that directing influence in his own research. I avoid the charge of making a personal attack on him by saying straight away that his own research has been of high quality, judged from that purely scientific standpoint which seems to me the proper basis of assessment. No one is ever justified in prophesying that any discovery will never be of material use, but there are few subjects which seem less likely, in the existing state of knowledge, to find material application than some of those chosen by Hogben for his own research. As examples one may choose his work on the development of the egg in gall-flies[68, 69], ants[68], and dragon-flies[70]. These are pieces of sound work, fitting into, expanding and illuminating our knowledge of the maturation of eggs, and more especially the behaviour of chromosomes. I should be happy if the work stood to my credit instead of Hogben's; but it was not inspired by material human need nor has it served material human welfare.

In this age, in which even certain scientists misunderstand the nature of science, one can scarcely lay heavy blame on politicians for falling into the same errors. Nevertheless there are limits to philistinism, which even a politician should not transgress. A good example of such transgression is afforded by Moore-Brabazon's remarks on lactation[103]: "The man who by his political efforts can get adequate milk to children deserves more of his fellow men than the inventor of the quantum theory; but in the narrow world of science, who gets the most attention and encouragement?" This remark is worth consideration, because it is an example of a fairly common attitude which should be exposed as false whenever it appears. I shall therefore analyse its falsity rather fully.

(1) By choosing milk and children as the subjects of his attack on science, Moore-Brabazon attempts to cause reason to be disturbed by emotion.

(2) Although cow's milk is good food, it is not a necessity

if the diet is varied. I have lived for two years among a well-nourished people who never drink milk except from their mothers' breasts. Many European children dislike milk after weaning and can easily be provided with alternative food-values in peace-time.

(3) The provision of milk supplies for those children who do not positively dislike it is good, but it is a matter which can be achieved by completely unoriginal people, as incapable of ever discovering anything as a politician. If those responsible for the provision of nourishment succeed in their work, they should be quietly commended for their ordinary efficiency. If they fail, they should be quietly replaced by others.

(4) It would be deliberate waste if original minds, gifted for research, were to undertake the job. If good scientists were to turn their attention to such matters, discovery would stop. What would be the present state of the health of the community (juvenile or adult) if men like Pasteur—an ardent believer in pure science—had been forced into the milk-distribution trade?

(5) If anyone says that the provision of milk for children is more important than the quantum theory, he must say also, to be consistent, that it is more important than the symphonies of Beethoven, the sculpture of Michelangelo or the writings of Shakespeare. Are we to become a nation of cultureless milkmen, because a few politicians cannot do a straightforward job requiring no spark of genius whatever?

(6) Planck's astonishing discovery that energy is not infinitely divisible is one of the most fundamental in science. It is difficult for the layman to comprehend, and seems contrary to "common sense". If the world is going to do without discoveries of this magnitude because politicians cannot appreciate them, science is going to stagnate.

(7) Moore-Brabazon calls the world of science "narrow".

Every fact about the universe which can be correlated with others into a system of demonstrable knowledge is the subject matter of science.

(8) Potential Plancks do not require any more "attention and encouragement" than Planck did himself when he was making his great discoveries, which he began to publish seventeen years before he received the Nobel Prize. Scientists are human beings and respond to appreciation shown by their colleagues, but they do not value the sort of "attention and encouragement" so much desired by politicians.

It is because loose talk like Moore-Brabazon's is so often directed against science nowadays, that this analysis has been undertaken. Most scientists regard it as beneath their dignity to reply, with the result that in the end science is actually threatened by uncomprehending philistines.

Those who have never experienced what Kropotkin called the "joy of scientific creation"[84] often regard love of science as something mystical and worthless. Actually these people, who fancy themselves as stern materialists, are much more mystical. If one carries their ideas to their logical conclusion, one arrives at nonsense. Everything, they pretend, must have material use. Everyone must concentrate wholly on nourishment, shelter, health and leisure: all else is useless. Nourishment, shelter, health and leisure thus become ends in themselves. They are not ends in themselves. People with the maximum amount of them are often the most miserable. To strive for them as though they were final ends is indeed mystical. Leisure particularly is misunderstood. By itself, with nothing to fill it, it is a positive evil, well illustrated in Sickert's painting, "Ennui", in the Tate Gallery. Never has leisure threatened so dangerously as now, when people are starting to demand passive amusement on the turning of a knob.

It is intolerable that some people should actually be

undernourished when there is enough food for all, but that can be put right by relatively few people without the whole population devoting every minute of life to thoughts of nothing except the material needs of man. Many people in all classes of the community who suffer great ill-health, even blindness or incurable disease, live full and worth-while lives just because to them nourishment, shelter, health and leisure are not ends in themselves. By concentrating on the real ends—art, music, literature, science, philosophy or the ethical part of religion—they find life worth living. Nothing is more useful in the final sense than the development of culture. Very few members of the community actually contribute to culture. Those who do so, do not want fame or wealth, but they want their work to go on. In the case of science, they have the special satisfaction, denied to musicians, of knowing that they not only serve the final ends of life but also build that secure and fruitful foundation on which technology can base itself to serve mankind in material ways. To misunderstand and therefore threaten science is to threaten both technology and culture.

Chapter 8

Science in the Age of Technology

WE live in an age in which one cannot fail to be impressed by the astounding achievements of technology. A concomitant of these achievements is the danger that we may unthinkingly claim to live in a particularly scientific age. In this chapter I am going to suggest various rather disconnected reasons for supposing that we should not lightly make this claim; but I want at the outset to make some remarks to avoid being misunderstood.

It may not be either possible or desirable to usher in an age in which everyone has a scientific mind. Literature and art would probably suffer, and there might be no benefit to various useful occupations. Even in subjects which are based on science, the scientific attitude is not absolutely necessary. An advanced medical student, entering not long ago an unfamiliar lecture-room at Oxford and seeing on the walls a number of drawings of rather obscure organisms unconnected with his own studies, remarked to a friend of mine, "A hell of a waste of time, learning about all these beasts". The "objects, materials and phenomena" of which Findlay wrote (see p. 84) were "a hell of a waste of time" to him. That need not make him a worse doctor. If I were ill, whether the illness were stomach-ache or cancer, I should not necessarily regard him as less able to give good advice than other doctors with a juster appreciation of the value of science. The full understanding of cancer, with everything that that means to sufferers from the disease, will of course only become possible when students of what this man would call a waste of time have pushed the basic sciences far enough for understanding to be possible; but existing scientific knowledge can be applied, and applied effectively, by those who could not possibly be scientists themselves.

It is probable that many people who have not the scientific attitude would grasp it readily and enjoy the fruits of it throughout their lives if facilities brought it within their reach. There are probably such people in every class and occupation, and it is my purpose to suggest in the next chapter some plans whereby a scientific culture might be spread among suitable people throughout the population. Before doing this, it is necessary to give a few random examples from various lines of evidence which seem to show that we do not live in a scientific age. Nowadays people assume that everyone is a sort of scientist, and sometimes even say so categorically[90]. People who will readily admit that composers or artists have somewhat different talents from the rest of the community are apt to regard anyone as presumptuous who claims a comparable difference for scientists. Many people are learning science nowadays and passing into the world as scientists—people to whom it would never have occurred to investigate anything if their parents, following the fashion, had not pushed them into the subject. These people, who know that they closely resemble other people, will naturally assume that other scientists are similar to themselves; and they will be the heartiest and most genuine opponents of the idea that there is such a thing as aptitude for research. There is a real danger that mass-produced scientists may eventually swamp the genuine article. If economic conditions or fashion were to make a hundred times as many people take up a musical career as happens today, many with no real musical ability would become composers. A lot of them would no doubt become proficient at such tasks as the arrangement of parts for a military band, but there would be no genuine inspiration.

The idea that we live in a scientific age is fostered artificially in various ways. We talk airily to our neighbour at a dinner-party about the second law of thermodynamics, but we only do this because we trust her not to let us down by asking whether we know what the first is. People talk freely and unaffectedly of ohms, volts and microfarads; but ask anyone who does so to find the resistance of a wire from the definition of an ohm, and see how he sets about it. An ohm is a thousand million times the resistance to the passage of an electric current shown by a wire which requires one erg of work each second to force through it a current which, flowing along a wire one centimetre long bent into a circular arc of radius one centimetre, exerts a force of one dyne on a unit magnetic pole placed at the centre of the circle, a unit magnetic pole being one that, when placed one centimetre from another unit pole, exerts on it a force of one dyne. I have not the faintest idea of how to begin, but I draw comfort from the statement[2] that it would take weeks of refined experiment by the most skilled observers to do it. Now the meaning of a volt depends on the meaning of an ohm, and is therefore more complex again, and to understand farads and microfarads one must understand not only the volt but also the coulomb. Of course we can all measure volts by the simplest possible means, but it is a very different matter to understand what the word means or the principles on which the calibration of the voltmeter depends.

It is not the bandying-about of scientific words which characterizes a scientific age. Russell[118] maintains that belief in reason reached its maximum in the 'sixties. Today there is evidence that about 40 per cent of the population of Britain has some degree of belief or interest in astrology. Harrisson[60], after studying hundreds of comments and conversations, says that "it is impossible to doubt that astrology is now a very considerable influence in determining the minor decisions of many private lives, and an appreciable contributory factor in influencing attitudes to wider, international events".

I was once at a public lecture on first aid, and the lecturer was explaining a method of artificial respiration. He stated that the operator should keep his fingers together. Immediately there was an outcry from the audience: people called out that the fingers should be kept apart. The lecturer now quoted a book of first aid in support of his argument: immediately members of the audience quoted another book in support of theirs. The course of lectures was not going to be followed by an examination, and there was therefore no set book to which it was profitable to adhere; but this did not affect people's behaviour. It now became apparent that some members of the audience had read the book quoted by the lecturer, and fierce arguments ensued as to whether the books said this or that. Chaos reigned for five or ten minutes. At last quiet was restored and the lecture started again, without anybody engaged in the argument having enquired how the separation or approximation of the fingers might influence the restoration of respiration, or suggested any way in which it could do so.

The day on which a student first starts to learn genetics, he is told about the inheritance of colour in Andalusian fowls. If Blue Andalusians are mated together, the progeny will tend to be in the proportion of one black to two blue to one speckled or "splashed" white, a close approximation to these proportions being obtained if a large number of chicks are raised. If, however, one of the blacks is mated with a splashed white, all the offspring will be of the desired "blue" colour (which is actually a slate-grey). The interpretation of these demonstrable facts forms a basis for the understanding of the principles of heredity. A special society[17], however, governs the affairs of the Blue Andalusian fowl, and the proceedings of such societies are based not upon science, but upon that particular variety of opposition to inductive methods which is called common sense. "It stands to reason", a breeder told a geneticist[31], "that if you continue to breed from the Andalusians you will ultimately fix the strain. It is common sense." Common sense dictated the destinies of the Blue Andalusians for fifty years. The black and splashed white progeny were killed or sold as "wasters"[123] and half the progeny only were Blue Andalusians. If the "wasters" had been kept and bred together, all the progeny would have been Blues; but this would not have satisfied the Poultry Club Council, which requires that the breed must have thoroughbred characteristics to come up to its Standard of Perfection.

When a discovery in pure science is reported in the popular press, the writer often either makes fun of it or feels bound to invent material ways in which mankind will benefit. It is assumed that readers will not be interested in science, except as a subject for fun or practical profit. Scientific films exhibited at public cinemas are usually accompanied by facetious commentaries.

Belief in a magical relation between whiteness and purity is strong enough to force the Government, at a time when national survival depends on the best use of food and money, to make elaborate arrangements to have the vitamins removed from wheat and subsequently returned to the flour. In putting forward deliberate propaganda it is possible to assume that a modern audience will not want to use reason at all. "Do you think for a moment", asked Quentin Reynolds[113] in a BBC broadcast, "that a man bearing the name of Winston Churchill will ever bend his knee to anyone named Schickelgruber?" This was not a private message to Hitler, but was broadcast at the time when the greatest number of British listeners would be listening. We are still in the age of magic.

It is not only among the general public that disregard of science and scientific method thrives in this technological age. I once knew a man who was interested in the fluctuations in abundance of a certain insect. He searched back in history for anything bearing on the subject. His research took him to ancient China, and to a time when war was in progress there. He tried to guess the effect of that war on the likelihood that the ancient Chinese would notice and record changes in the numbers of the insect. He wanted to use his guess in a scientific account of periodical fluctuations in abundance.

Those who have not studied the history of science are apt to think that apart from a few geniuses whose names are household words, there was little science till a hundred years ago. A fleeting glance through the two centuries before 1840 will show how wrong this impression is. Even if one excludes the most familiar names, restricts oneself to physics and merely dips casually here and there into the subject, it will not be found possible to retain the idea that science was sleeping in those two hundred years. At the beginning of that period von Guericke was experimenting with vacua and air-pumps. Later in the seventeeth century Mariotte discovered the law of the compressibility of gases independently of Boyle. During the first half of the next century Cuneus discovered the electrical principle which led to the Leyden jar, and in the process of doing so got such an electric shock—the first one, apart from lightning, ever felt by man —that he announced that he "would not, for the crown of

France, expose himself to a second such shock"[2]. The very
next year Bishop Watson, at Shooters Hill, near London,
was sending an electric shock from a Leyden jar through
two miles of wire supported by wooden poles. In 1757
Dollond invented achromatic lenses and made possible the
wonderful optical instruments of today. About the turn of
the century Monge, accompanying Napoleon on his Egyp-
tian campaign, explained the phenomenon of mirage, and
Chladni investigated the vibration of strings, rods and plates
and laid the foundation of modern acoustics. Malus discovered in 1809 that light is polarized by reflection. A dozen
years later Seebeck discovered how an electric current may
be produced by heating the junction of two wires made of
different materials, and thus originated the thermopile. In
1827 Colladon and Sturm were measuring the velocity of
sound in water by experiments made with the aid of two
boats moored a known distance apart in the lake of Geneva.
In 1834 Lenz announced his law relating the direction of an
induced electric current to the movement which induced it.

If an authority were to pursue the history of physics in
those two centuries, omitting all the work done by the most
famous men, a volume could be filled; and if all sciences
were included, many volumes. A glimpse has been enough,
but another view-point shows that science was wide awake
during those two centuries: the great men discovered the
same things about the same time. Charles discovered in
1786 the law of the expansion of gases when heated which
still bears his name. He omitted to make his discovery
known, and Dalton rediscovered it in 1801 and Gay-Lussac
the next year.

Anyone who cares to look at the history of scientific
thought will find that right through the centuries there were
people groping towards the attempt to obtain knowledge by
observation, experiment and reason. Hippocrates, and a

little later Aristotle, were already developing the method of science about twenty-three centuries ago, the former in pathology and the latter in biology. How much Aristotle owed to Hippocrates, it is impossible to say: he only refers to him once in his writings[12]. From that time to the present day it has been open to people to use the scientific method, and it is perplexing to remember that despite this the great majority of the people of all countries have always preferred to be guided by tradition, authority, desire or magic. Hippocrates and Aristotle had to think out the method of science for themselves, and the other great scientific philosophers of past ages had little to encourage them in what they read. Today there is a profusion of books on science, but the effect in producing a scientific culture is small. People "do stinks" at school; they "make" chlorine or dissect an earthworm: but there is little realization of the great principles which, in the hands of an infinitesimal fraction of the human race, have made possible the building up of such a vast body of demonstrable knowledge.

Religion has probably played a part in preventing the growth of scientific culture, for in every religion except Unitarianism the use of reason is only accepted with certain safeguards. It must be said at once that great scientists, from Pasteur downwards, have not rarely been religious, but nevertheless formal religion has probably been a force acting in opposition to a wide understanding of the methods of science. Fascism, Nazism and Communism, which have something in common with religious movements, are also a detrimental force because they put other values above truth. In science there is no loyalty to creed, party or class: only to demonstrable truth and reason.

Chapter 9

People as Scientists

"He who has once in his life experienced this joy of scientific creation will never forget it; he will be longing to renew it; and he cannot but feel with pain that this sort of happiness is the lot of so few of us, while so many could also live through it—on a small or on a grand scale—if scientific methods and leisure were not limited to a handful of men."   KROPOTKIN[84]

NOT everyone's mind provides a fertile soil for the growth of science, but the most unexpected people sometimes respond surprisingly to the sowing of a seed. Go for a country walk with a non-scientific friend and suggest tentatively that the hills and valleys are not arranged haphazard, but result from the geological history of the world in ancient times. You are likely to be deluged with questions, and unless you happen to be an authority on the local geology, you may find yourself quite at a loss to answer many of them. The intelligent ignorant person always asks the best and hardest questions, because his mind has not been taught to follow a rut.

A wider spread of scientific culture would not only bring happiness to many who are not and cannot become professional scientists: it would benefit the professional scientists themselves. Indeed, one can scarcely speak of a culture when the activity is confined to those who make their living from it. A cultured public is a necessity for good literature and music. It is valuable also for pictorial art, though here the relation between what might be called producer and consumer is much less satisfactory. People may seek to make

money by attaching a rarity-value to the works of a certain
artist, or falsely boost his virtues for the same purpose, so
that a carefully fostered deceit may vitiate public taste. A
living artist may spoil the rarity-value of his works by
painting more pictures: he is better dead. An unfortunate
painter may even see it mentioned in print[139] that his work
fetches more per square inch than that of any other painter
has ever fetched during the painter's lifetime. Luckily such
degradation of culture can never afflict the scientist any
more than the writer or musician, for the work of all three
can be printed and reprinted indefinitely without detriment,
and there is no scope for the deliberate attachment of false
values. The scientist is generally appreciative of appreciation, if it is based upon understanding: fame cannot help
him. If there were a more widespread scientific culture,
professional and amateur scientists could render one another
the sort of mutual help given by composer and musical
public. Just as amateur musicians play music, so also
amateur scientists can actively pursue science; but whereas
an amateur musician is unlikely to add anything permanent
to music beyond his own contribution to the spread of
musical culture, an amateur scientist can make genuine discoveries, and these will be accepted into the general body
of knowledge and become the foundations for further
research.

Amateur science is regarded with some disfavour, even
contempt, today. It was not always so. Einstein and Infeld[43]
have remarked that "nearly all the fundamental work concerned with the nature of heat" was done by amateur
physicists. People tend to forget that one of the greatest
scientists of all times, Charles Darwin, was an amateur. He
was, of course, a fairly rich man. Rich men are much rarer
nowadays, and no one would wish to suggest a culture
specially designed for them. Wealth, however, is by no

means a necessity for the amateur scientist. Anyone who thinks it is should read the life-story of Hugh Miller[102], the quarryman, stone-mason and accountant, who was a well-known geologist and palaeontologist and author of "The Old Red Sandstone"[101]. A shop-keeper of Thurso helped to furnish him with the specimens used in his work. Miller wrote enthusiastically of another amateur scientist as "one of the best . . . geologists in Forfarshire"[102]. That would not be a likely description of anyone today. The words were written in the second half of the nineteenth century, when science probably progressed more rapidly than ever before or since, and was widely valued as an end in itself.

What can be done to stimulate a wider interest in science today?

It may be suggested that the approach is wrong. Nothing could be more unlikely to produce a scientist than the knowledge that one must "do" certain sciences for the school certificate. The first requirement is that the young boy or girl should get to know something of the scope and method of science. These subjects will probably be found harder to teach than the bare facts of a science, because the teacher may know less about them. The teaching, however, need not be elaborate. The principles of observation, experiment and inductive reasoning are not complex, and quite a small child can be made to understand the difference between objective and subjective experience. No great knowledge of the various sciences is required if one is only going to say what sorts of subjects are covered by each. It will, of course, be necessary to show that the scientist cares nothing for authority or tradition, and it may not be found quite easy to reconcile this, in the untrained minds of the pupils, with the respect properly shown by scientists for the great discoverers of all times. The life-stories of the greatest discoverers provide a human background which may give an interest in science to those children who find the plain facts rather dry.

If the object of the teacher is to find out which of the pupils is the best adapted for further training in science, the simplest work is the best. It is no good to try to rouse a false interest by giving a necessarily over-simplified account of some wonderful technological achievement like wireless telephony. It would be far better to teach a child how to weigh as accurately as the available apparatus will allow, or to measure or count or compare familiar natural objects. Elementary schools cannot afford to purchase elaborate scientific apparatus, but it may be suggested that quite good balances and microscopes might be circulated from school to school, so that there would be an occasional opportunity for each child to work with real instruments under the supervision of the teacher. It should not be hard to spot the child who delights in deftness and accuracy and is worth starting on a scientific career. At present rich people have a much greater possibility of interesting their children in science than poor, and there is probably an untapped wealth of talent in all classes which needs a little encouragement if it is to be found. It is true that a Faraday will rise from any circumstances, but a great part of science is the work of much lesser but very good men, who would not have been scientists at all if their parents had not been fairly rich.

The waste of talent by non-recognition is not overcome by the scholarship system. If talent in science announced itself as early and unmistakably as musical talent, the scholarship system would suffice; but there is a tendency for this system to favour the precocious, and scientists are not particularly precocious. When Bateson, the eminent geneticist, was seventeen years old, the headmaster of his school wrote of him, "it is very doubtful whether so vague and aimless a boy will profit by University life"[14]. No one

thought Charles Darwin clever as a child, boy or young man.
J. J. Lister, who invented the principle on which all modern
high-power object-glasses for microscopes are constructed,
did not make any discovery until he was thirty-eight years
old[86a]. Success as a discoverer probably depends much
more upon inclination and perseverance than on unusual
cleverness in childhood. With reasonably promising material,
intelligence will develop as the necessity for it arises.

The wider and more unspecialized the teaching of science
in schools can be, the better, provided that there is no
accompaniment of inaccuracy. If a boy or girl in the 'teens
shows special enthusiasm for a limited field, as often happens, the development of that field can well be left to out-
of-school hours. The product of the school should not be a
prodigy of learning in one science, ignorant of the methods
and scope of science as a whole and its position in human
culture.

When I was a boy at school, I was spending one half-
holiday afternoon examining the eye of a beetle with my
microscope. No biology was taught: indeed, it was a rare
subject in schools at that time. The headmaster chanced to
pass by and asked me what I was doing. The investigation
was not approved. I was told that I was wasting my time and
was sent off forthwith to spend my half-holiday doing some-
ting useful, mowing the lawn. I was determined to study
zoology and devoted every available moment to it despite
the lack of encouragement at school. At last my real opportunity came, at Oxford. The Warden of my College sent for
me and asked me what subject I wanted to study. "Zoology",
I replied. "Would it not be better", asked the Warden, "to
study something useful, such as medicine?" I was too
inexperienced and shy to explain to the man who had taken
a triple first class (in classics, mathematics and theology)
that medicine stands on the basic sciences like a house on its

foundations and that science has a value as an end in itself; but I studied zoology.

Some amount of discouragement in the later years of a boy's or girl's development is probably not a bad thing: it will show up the enthusiast. It would be sad to see the research laboratories of pure science cluttered up with non-enthusiasts. J. G. Crowther[26] has stated that the majority of British scientists would probably be glad to resign from research permanently, in exchange for high administrative positions. I can only say that I cannot disprove this statement statistically, but it is absolutely contrary to my experience of research scientists, whose enthusiasm for their work is only comparable to that of the true musician or artist for his. If there were too much encouragement the wrong people might be led astray into thinking they had the gifts of an investigator. That great scientist, Banting, discoverer of insulin, believed that people should only take up research if they had an impelling urge to do so. "Do not enter upon research", he told the students of Edinburgh University[22], "unless you cannot help it." This is memorable advice.

Of those who have become fascinated with science as children, some will pass on to a scientific career and others will take up other occupations and retain their interest. There are many people in other occupations today who are interested in science, though probably less than in the second half of the last century. What can be done to let them take a fuller share in scientific culture?

There are certain sciences which are particularly suited to the amateur. Almost no research in chemistry is carried out today by amateurs, perhaps none at all. I do not believe that amateur chemistry is a modern impossibility, but clearly other fields are more inviting, especially those subjects which, as has been shown in an earlier chapter, may be conveniently grouped together as natural history.

It is in unexplored country that many branches of natural history can best be studied; but exploration cannot be for everyone, and nature reserves are nearly as good for many purposes. The nature reserves of North America are magnificent examples of what can be done by freeing a small fraction of a country from commercial exploitation in the interests of science and recreation. In Europe there are such fine reserves as the Swiss National Park near Zernez in the Engadine[18], where natural scenery, flora and fauna are preserved for enjoyment and study by the people of today and by posterity. The National Trust has made a start in Great Britain. It is greatly to be hoped that after the war more national nature reserves may be started in various parts of Great Britain, for the use of professional and amateur scientists and of the general public. Some could be quite small, perhaps only a few acres in extent in certain carefully chosen places; others covering wide areas of country of little commercial value. In each a small central part should be set aside and no one allowed to enter except serious students, amateur or professional.

The scientific study of natural environments is in its infancy. The evolution of plants and animals took place in natural environments, and evolution can never be understood unless such places are set aside for study. Not only in industrial countries but also through most of the world man is effacing natural vegetation and placing beyond the hope of study those complex interrelations of plants and animals which have resulted in the evolution of living organisms as we know them today[65]. What is wanted is not the artificial protection of a few selected rare species in special sanctuaries, but real nature reserves where the interrelations of organisms may be exhibited without human interference. The extinction of rare species is regrettable, but probably it does not generally have much effect on the flora and fauna as a whole. The introduction of a few exotic species may, on the contrary, completely upset the natural conditions of life. Nature reserves will be useless in a country if the deliberate setting free of species introduced from abroad is permitted, for they will be likely to spread to the reserve and it may be found impossible to eradicate them. Great Britain is rather fortunate in this respect. Not many people have followed the bad example of the Duke of Bedford, the chief introducer of the grey squirrel, whose natural home is the central and eastern parts of the United States and south-eastern Canada[96, 97].

Other parts of the world have suffered heavily. One thinks with despair of the man who got a medal for introducing rabbits into Australia, and of all those, undecorated but assiduous, who wrecked the intensely interesting fauna of New Zealand by introductions. A walk through that country now reveals a preponderance of introduced birds. No other place in the world had such an extraordinary bird-fauna as the Hawaiian Islands. I well remember stepping ashore there optimistically, quite forgetting that the vandal had been on the spot long before. If a mad millionaire decided to spend his fortune in destroying Wallace's line by transporting Oriental species to the east of it and Australian to the west, his error would differ in degree but not in kind from that of a person who deliberately and capriciously tries to naturalize and set free any exotic species anywhere.

It is sometimes argued that the introduction of an exotic species is natural because it occurs; that it is an experiment, and therefore interesting. Suppose that the charwoman who cleans a chemistry professor's laboratory thinks it fun to mix the contents of two of his test-tubes. Let us imagine that when he expostulates, she tries the argument on him. "Everything that has happened is natural", she says; "the contents of the two tubes have reacted in natural ways, and you ought to be interested in studying the reactions." I shall not elaborate the professor's answer; but in a cooler moment he will reflect that the only possible scientific study of the incident would be a psychological investigation of the caprice which made her interfere so thoughtlessly and catastrophically with his research. The biologist has often cause to think these thoughts.

There is every degree of interference with natural habitats from the breaking of a twig as one pushes one's way through an unexplored tropical forest to the conversion of country-side into a city. Perfectly natural conditions, in the sense that they are absolutely unaffected by the will of man, can seldom be provided in a nature reserve, but the best is the enemy of the good, and very good nature reserves could be set aside if counsels of perfection were not allowed to bar the way. A sharp distinction must be drawn, nevertheless, between a true nature reserve and old-fashioned country-side. The scientist, as such, has no particular anxiety to preserve a certain stage in the evolution of agriculture because it is pleasant to look at old farmsteads and their surroundings. It certainly is pleasant, but it is not science.

Today it has become especially desirable that there should be a re-awakening of interest in natural history. In the growing science of ecology[44, 45] there can well be co-opera-tion between professional and amateur scientists, to the benefit of both. The British Trust for Ornithology is already encouraging such co-operation in its own field. Genetics, long thought to be a science of the laboratory, garden, and animal breeding establishment, is not going to remain for ever so tightly constricted. E. B. Ford, the originator of a movement for the study of genetics in natural habitats, has already made important studies of the actual occurrence of evolution in our own times[49], and has not hesitated to collaborate with amateur naturalists. Studies of this kind could probably be multiplied and would result in great advances in knowledge. Everywhere problems seek solution: we are ignorant of much that concerns the behaviour of the very commonest animals. An amateur naturalist, Eliot Howard, is one of the foremost authorities on the behaviour of wild animals and in particular on territory in bird-life and the significance of bird-song[74, 75]

The non-professional scientist must always strive to keep his interests as wide and deep as possible. Merely to know the names of organisms in a limited group is scarcely more scientific than stamp-collecting. Private collections are generally useless; collections should only be made for museums so that the specimens will be available to others. It is far better to observe than to collect. I have seen a printed discussion of whether the collector of birds' eggs should know anything about what lies within the shell. Anyone who limits himself to the study of the shell ignores the significance of what he is studying, and scarcely ever does the collector even study the shell itself, its minute structure, the chemistry of its pigments, or the methods of its formation.

Many naturalists are extraordinarily limited in their outlook, without realizing the fact. Some of them, intensely interested in the outsides of birds, are surprised if they discover that a professional zoologist is ignorant of some feature of the coloration or habits of a bird with which they are familiar. It is as though a man devoted his life to the study of the external appearance of green bridges, and was astonished, on discussion of his favourite theme with a professional engineer, to discover that the latter was totally ignorant of the existence of a certain green bridge in Venezuela. Fisher's "Birds as animals"[48] provides a healthy corrective to the narrow outlook of some students of bird life. (It may be remarked that the title which he chose suggested those which I have adopted for two chapters of this book.)

Amateur naturalists need not all concern themselves with problems of great complexity. A little story will illustrate this.

A trader on the coast of Labrador, George Cartwright, brought some rabbits with him from England. On 24th April 1778, he wrote in his diary an account of a doe giving birth to young in his dining-room. He made observations on the development of the young, and recorded that on the eleventh day they "began to see". His diary was subsequently published[21].

More than a century and a half have gone past since George Cartwright observed his rabbits. Ask any zoologist today at what age the eyes of the young rabbit open. The matter is not entirely insignificant, because the rabbit is so much used as a type in teaching zoology. The zoologist will at once refer you to Barrett-Hamilton's account of British mammals[13]. This author, however, got his information from a footnote in a book by Harting published in 1898[62]. Although Harting does not give a reference, he must have been quoting from a book by Daniel published in 1801[29], for every particular length of pregnancy, age when ears can be moved, age when eyes open, and age when ears can be erected, are exactly the same. Daniel, however, never claimed to be the original observer: he got his information straight from Cartwright. So we are back again in Cartwright's dining-room in Labrador in 1778, apparently the place of origin of the best modern information on the age at which the rabbit's eyes open. The information is based on the study of a single litter of tame specimens in an extraordinary environment. One is reminded of the scholars of long ago, who sought to settle once and for all the vexed question of whether mares have canine teeth, not by looking to see, but by reference to the works of Aristotle. Lest anyone should hurry off to

I

study the eye-opening of young rabbits, I must give three warnings. First, the practical difficulties are not as small as might be thought; secondly, statistical methods must be used; and thirdly, one should make sure that some tedious German has not already found the mean age of eye-opening to the nearest minute.

To encourage the growth of scientific culture a new journal and a new society seem necessary. There is no British journal of general science suitable for reading by amateurs. "Nature" is now the only general scientific journal published in Britain.* Every scientist will admit its usefulness, but it contains so much that is incomprehensible outside one's own province of knowledge, that it can scarcely claim to present science as a whole to either professional or amateur. Further, it has become increasingly interested in technology, presumably because it is read by more technologists than scientists. The very name of the journal has thus tended to lose something of its meaning. Although the extract from Wordsworth is still printed on the cover every week,

> "To the solid ground
> Of nature trusts the Mind that builds for aye,"

yet one imagines that a good many minds which eagerly devour its pages are building for the immediate future and trusting more to business than to Nature. It was by a curious chance that at the beginning of 1935 Nature lost her capital letter in the little verse (while mind acquired one); for it is since about that time that the fashion has arisen to confound science with technology and to overstress practical applications at the expense of basic knowledge.

A journal of scientific culture would be readable throughout by any professional or amateur scientist, and would keep

* The *School Science Review* is excellent, but addressed to a special group of readers.

the interests of both wide and lively. It would not be impossible to write about science in a generally intelligible way, for there are excellent examples of fairly recent books on scientific subjects which the general reader can study with interest and which are profitable also to scientists working in different sciences: for instance, Jeans on acoustics[82], Findlay on chemistry[47], Parsons on biochemistry[107], Skene on British plants[121], Ford on genetics[50], Tansley on psychology[126], and Trueman on the geological basis of scenery[135]. Darwin wrote particularly intelligibly and the public seized what he wrote with avidity; every copy of the first edition of "The Origin of Species" and 5,267 copies of "The Expression of the Emotions in man and animals" were sold on the respective days of publication[36]. In this twentieth century there are branches of science in which new advances in knowledge can be published in quite an intelligible form: one may instance Elton's "Animal Ecology and Evolution"[44], Fraser Darling's "Bird Flocks and the Breeding Cycle"[30] and Eliot Howard's "Territory in Bird Life"[74].

The books that I have mentioned show that a general journal of scientific culture is a possibility, and one may earnestly hope that after the war such a journal will be founded. One may hope also for the formation of a society for unprofessional science, to help amateurs in all matters connected with their interests, and especially in arranging for mutual help between them and professional scientists. The Fellowship of the Royal Society is the highest honour that is conferred on scientists in this country, and there is every reason to believe that the existence of this special honour is a stimulus to many research workers. It is very difficult nowadays for an amateur to make such a large contribution to knowledge as to gain this Fellowship, but the principle of formal recognition of good research could be extended to the proposed new society, which might confer Fellowship for investigations of particular merit, while allowing membership to all with the interests of science at heart. It is unfortunate that the amateur scientist should so often go without proper recognition, while people lacking the true spirit of science get themselves labelled with degrees.

There are many ways in which the living culture of science could be encouraged, and only a few can be briefly touched on here. One last suggestion may be put forward. People with active brains who enter a scientific museum are overwhelmed by the superabundance of what is put before them: they do not know which way to look: a few cubic feet of space would occupy hours of study, and from every side other exhibits call with equal insistence. It is as though one entered a gigantic lecture-room in which all the lectures of a three-year course were being delivered simultaneously. Is it not possible that public museums are wrongly arranged? Might not the greater part of the public space be devoted to a few exhibits, and these constantly changed and the changes announced in the local press? The number of exhibits would be as low as was compatible with everyone getting a good view. Behind the scenes there would be the same wealth of material as before, readily available to the serious student, amateur or professional. This scheme would stop the aimless wandering about that one notices in museums, and the would-be wanderer might leave the building with a few definite new thoughts in his head, instead of a haze of ill-digested information.

The suggestions put forward in this chapter for the development of a wider and deeper scientific culture are crude and tentative. I hope that others, aware of their crudity and more far-seeing than I, but no less anxious that the culture of science should prosper, will put forward better ideas and transform them into realities.

Chapter 10

Prospect

THIS book has been written in the hope that it may exert a little influence on people's attitude to science during the period of reconstruction after the war. So many books have been written lately by those who equate science with technology and wish that research should be centrally planned, that the layman might easily be forgiven for thinking that these books represent the views of the majority of scientists. Unfortunately the majority say nothing and let their case go by default. Books and speeches pour forth from the "planners". The British Association for the Advancement of Science organizes a so-called "conference" at which all the leaders of the "planning" party deliver addresses. During the whole of six sessions lasting three days, no one is allowed to speak except those chosen in advance by the organizers, and none of the leaders of the movement for freedom in science is chosen. A "Charter" is produced by the Association, which includes the words, "The basic principles of science . . . are influenced by the progressive needs of humanity"[57],* as though the search for objective truth were futile. "Nature" moves with what its editors think to be the tide. "The Scientific Worker" adds its influence.

Still the great majority of scientists make no public

* Since the holding of the conference the Association has rewritten the sentence, but the fact that it could be solemnly pronounced by the President and then published shows to what depths a body supposed to represent science may sink.

utterance. "It can't happen here", they say in private; "leave us alone; we only want to get on with our research." Exactly; so do I. It is because I want scientists to be free to do their own research that I have written this book. They will not be free if they do nothing and let those who would dictate their research for them do all the talking and writing. Only one little book has appeared in opposition to the planners' library. Polanyi's "The Contempt of Freedom"[111] contains a masterly short statement of the case for freedom in science.*

There is only one aspect of politics which directly concerns scientists as such. When their freedom to pursue their own research is at stake, they are directly concerned, and they are in a position to form sounder opinions on that subject than anybody else. They should make their opinions heard. In other matters it is often best for them to leave politics alone. A research worker cannot obtain the quietude necessary for the fulfilment of his duty—that of discovering demonstrable truth—if he throws himself into the stress of party politics. If he goes so far as to use scientific prestige to get political power, he prostitutes his talent. Hopkins, a pioneer of vitamin research and a practical benefactor of humanity, said that it is impossible not to sympathize with the view, commonly held by scientists, that they will be more useful if they continue their chosen work in its proper environment, than if they give up time to the question of the social implications of their calling[73].

The planners adopt a pharisaical air of ethical superiority. They presume to give the impression that they alone are concerned with the welfare of humanity, while other scientists selfishly study nothing but their own inclinations and convenience. Since the planners raise the question of

* See also Polanyi's "The growth of thought in society" (*Economica*, Nov., 1941).

ethics, they must be answered. Everyone, scientist or not, has the opportunity every day of his life to promote the welfare of others in inconspicuous ways. The scientist who quietly and unostentatiously does good to others by simple kindness, thoughtfulness and justice is more likely to be a useful member of the community than one who loudly boasts his social conscience and his determination to improve other people's lot by seeing that it is dictated to. them. Whereas political stress dulls the scientific mind and makes it less capable of originating new ideas, the simple un-advertised virtues have no such effect. Simple, good, hate-free men like Charles Darwin and Michael Faraday are inspiring examples to all who come in contact with them or who read the stories of their lives.

One hopes that the period of reconstruction will provide a fairer share of the good things of the world for everyone and more equal opportunities for each person to serve the community in the way for which he happens to be best fitted. More equal opportunities for members of all classes to live the scientific life would give great promise for the advance of science. Great scientists have arisen from all classes. Even if it be true that genius will always manifest itself, yet, as we have seen, an enormous contribution to scientific research is made by talented people who are much less than geniuses. Such people can best be recognized and secured by giving equal opportunity to all.

The coming of a real democracy, then, would bring a great hope for science. It would also bring a danger. That danger must be faced, though I am aware that what I am about to say on this subject lies patently open to miscon-struction and partial quotation by those who do not want to face the facts.

Few superstitions have less scientific foundation than snobbery. That anyone should claim special privileges because he was born in a certain class is ridiculous. If in fact he is innately superior, then his superiority should manifest itself without the necessity for privilege. Half a century ago, snobs fawned upon the aristocracy and the richer classes of the community. Scientific analysis of the social distribution of innate intelligence suggests that the professional classes are on the average somewhat superior in this respect[11, 21a, 21b], but there is so much overlap between all classes that the scientific basis of professional-class snobbery is small, so far as innate qualities are concerned. Today snobbery rears its ugly head again, but now it turns it in the opposite direction, and the tendency is to fawn upon the factory-workers. No scientific evidence supports this kind of snobbery, which is just as baseless as the others. Harrisson has done well to expose its falsity[61]. To try to obtain advantage or security for oneself by falsely attributing special innate virtue to any class is contemptible.

In the professional class there is often a tradition of respect for certain ends-in-themselves, such as art, music and literature, and so the children tend to grow up with a feeling that such ends have a primary human value. If the child of a factory-worker were to grow up in the same surroundings, he would undoubtedly tend to respect those ends in the same way. Actually he does not generally grow up in surroundings where culture is much respected (though of course there are splendid exceptions). That is the fault neither of the parent nor of the child, but it is a fact; and it has resulted in the adults of the poorer people generally having less respect for culture than the members of the professional classes.

Scientific culture has certain advantages not shared by artistic and literary culture. When the historians of the distant future look back on the pitiful struggles of man in the years between the two great wars, they will note with sympathetic understanding that many of the artists, composers and writers were jolted by their environment to produce work which suited it but was without lasting value. They will know the reason for the child-like and schizophrenic paintings, grotesque statues, cacophonous music and chaotic literature of the period. They will recognize that the scientists of the time, in contrast, went solidly ahead—more slowly, probably, than in the quieter years before, yet straight ahead—to contribute, as always, to the eternal value of truth. Because science is uninfluenced, except in speed of progress, by passing phases in the external world, because even an environment that tends to shatter artists and composers and writers cannot throw science off its balance, science may perhaps claim to be a unique manifestation of the human mind. It progresses more certainly than other cultures because its progress is more necessarily additive. Its potentialities are unbounded, for each generation starts where the one before left off. Only two things can kill scientific progress. Those two are "planning" and the confusion of science with technology (a confusion which is comparable to finding nothing in music but box-office receipts, musicians' salaries and a stimulus to the martial spirit, and which would in the long run be as damaging to technology as to science).

If a country suddenly puts the government into the hands of the poorer members of the community, culture of all kinds is genuinely threatened unless it has already spread throughout the population. The Russian revolution provides an instance. Interest in the ballet pervaded all classes in the big cities before the revolution, and ballet survived and still flourishes. Science, never very strong and a closed book to the poor, was struck a staggering blow. Although vast sums are said to have been spent on Soviet science, the return (in science as distinct from technology) has been poor. A few

British scientists take every opportunity to praise Soviet science. Haldane[59] feels it necessary to tell us that certain Soviet scientists have not lost their jobs, so far as he knows, although they hold views (held almost universally by the geneticists of the rest of the world) which conflict with views expressed by Engels. We are supposed to admire the toleration which admits discreet opposition to the sacred writings. Even Haldane has to admit that the genetical view favoured by Soviet authority "appears to be untrue in the light of actual biological research". One cannot fail to ask oneself whether the same people who praise Soviet science so loudly would have bothered to do so or to make such careful excuses for its shortcomings[16, 26], if the same scientific output had come from a non-communist state. The praise springs from political, not scientific enthusiasm. It must be remembered that the making of many guns, tanks and aeroplanes is a technological achievement, which has no connexion with the advance of scientific research.

At the time of the French revolution, the president of the court which sentenced one of the greatest chemists of all time to death remarked, "The republic has no need for scholars"[94]. (Lavoisier was executed the same day, though guiltless of any crime.) One hundred and forty-seven years later the Soviet ambassador told the British Association for the Advancement of Science (on 27th September 1941), "We in the Soviet Union never believed in pure science." If the poorer people in this country were to determine its internal policy without accepting advice from others, culture of all kinds would for a time be endangered. The phrase scientific culture, or science as an end in itself, would not be in the least understood. Technology only would be accepted. Science would be dealt a heavy blow, from which technology would eventually suffer as heavily as science itself. A warning from scientists is urgently necessary, and instead of the warning there comes a steady flow of propaganda in the wrong direction, urging the smothering of science under technology and the substitution of planning for free enquiry. A large part of the propaganda is poured out by those who have themselves been scientific research workers, which would be hard to understand if we could not find clear evidence of political emotion quite obscuring clarity of scientific vision.

If people who have had the good fortune to grow up in an environment in which culture is respected would give the warning in time, they would serve the community well. If only snobbery could be eliminated, culture of all kinds spread throughout the community, and suitable talented people taken freely from all classes to enrich all professions, civilization might progress as never before in the history of the world. Everyone in a true democracy would have the opportunity to live a full life according to his innate capacities. Those gifted people who found themselves attracted towards immediately practical pursuits would find an absorbing life-work in technology. Those of a less practical disposition would serve the community equally well as scientific research workers or creators of culture in some other sphere. Music, art, literature, philosophy and science would be widely appreciated among those not talented enough to be creators themselves. I do not agree with the pessimism of Raymond Mortimer[106], when he says that he sees no grounds for the assumption that the great majority of human beings are potentially capable of enjoying the arduous exercise of their intellects and æsthetic sensibilities. The great majority of children, I think, are capable of being trained to appreciate some sort of culture.

True appreciation, based on real understanding and widely spread through the community, would react on the creators to inspire further creative effort. With free enquiry, free criticism, lively discussion and the opportunity given to all to understand and enjoy the great gifts that culture can bring, a genuine democracy would produce a flourishing civilization in which each person would be given the chance to feel that life was full and worth-while and exciting. Widely diverse in tastes and interests and talent, each man and woman could find a satisfying niche.

Another and a vastly different prospect looms ahead. An ugly new god called the state demands worship. Nourishment, shelter, health and leisure are falsely regarded as ends in themselves. Culture is looked down on with contempt. Science is equated with technology and both decay. Individualism and free enquiry are ridiculed. Everything is planned from "above". A dreary uniformity descends. Each person is a cog in a vast machine, grinding towards ends lacking all higher human values.

It is for you to choose.

List of References

(1) ADAIR, F. E. 1931. *Cancer*. Lippincott, London.
(2) ALDOUS, J. C. P. 1910. *An elementary course of physics*. Macmillan, London.
(3) ALEXANDER, S. 1933. *Beauty and other forms of value*. Macmillan, London.
(4) ALLMAN, G. J. 1880. "On 'Limnocodium victoria', a hydroid medusa of fresh water." *Nature*, 22, p. 178.
(5) ALLPORT, N. L. 1936. "*p*-Aminobenzenesulphonamide." *Quart. Journ. Pharm.* 9, p. 560.
(6) ANDRADE, E. N. DA C. 1938. "Science in the seventeenth century." *Nature*, 142, p. 19.
(7) ANONYMOUS. 1908. Article on Cavendish in *The illustrated Chambers's Encyclopaedia*. Chambers, London.
(8) ARMSTRONG, E. F. 1941. "Henry E. Armstrong." *Nature*, 147, p. 373.
(9) BACON, F. 1605. *The advancement of learning*. Reprinted in Everyman's Library. Dent, London. (For Bacon see also Verulam.)
(10) BAKER, J. R. 1940. "Science in the U.S.S.R." *New Statesman and Nation*, 2nd March, p. 276.
(11) BAKER, J. R. and J. B. S. HALDANE. 1934. *Biology in everyday life*. Allen and Unwin, London.
(12) BARKER, E. 1929. Article on Aristotle in the *British Encyclopaedia*, 14th edition. London.
(13) BARRETT-HAMILTON, G. E. H. 1913. *A history of British mammals*. Gurney and Jackson, London.
(14) BATESON, B. 1928. *William Bateson, naturalist: his essays and addresses, together with a short account of his life*. Cambridge University Press.
(15) BERNAL, J. D. 1939. *The social function of science*. Routledge, London.
(16) BERNAL, J. D. 1941. "Present day science and technology in the U.S.S.R." *Nature*, 148, p. 360.
(17) BROOMHEAD, W. W. (No date.) *Poultry breeding and management*. New Era Publishing Co., London.

(18) BRUNIES, S. 1920. *Der schweizerische Nationalpark.* Schwabe, Basel.

(19) BUCKLEY, H. 1927. *A short history of physics.* Methuen, London.

(20) CAJORI, F. 1899. *A history of physics in its elementary branches.* Macmillan, London.

(21) CARTRIGHT, G. 1792. *A journal of transactions and events, during a residence of nearly sixteen years on the coast of Labrador.* 3 vols. Allin and Ridge, Newark.

(21*a*) CATTELL, R. B. 1934. "Occupational norms of intelligence, and the standardization of an adult intelligence test." *Brit. Journ. Psychol,* 25, p. 1.

(21*b*) CATTELL, R. B. 1937. "Some further relations between intelligence, fertility and socio-economic factors." *Eug. Rev.,* 29, p. 171.

(22) COLLIP, J. B. 1941. "Frederick Grant Banting, discoverer of insulin." *Sci. Monthly,* May, p. 473.

(23) COWDRY, E. V. 1938. *A textbook of histology: functional significance of cells and intercellular substances.* Kimpton, London.

(24) CROOKES, W. See Gregory (55).

(25) CROWTHER, J. A. 1941. "Science and government." *Nature,* 147, p. 415.

(26) CROWTHER, J. G. 1941. *The social relations of science.* Macmillan, London.

(27) CUVIER, LE CH^ER. 1817. *Le règne animale distribué d'après son organization.* 4 vols. Deterville, Paris.

(28) DALCQ, A. M. 1938. *Form and causality in early development.* Cambridge University Press.

(29) DANIEL, W. B. 1801. *Rural sports.* Bunny and Gold, London.

(30) DARLING, F. F. 1938. *Bird flocks and the breeding cycle: a contribution to the study of avian sociality.* Cambridge University Press.

(31) DARBISHIRE, A. D. 1911. *Breeding and the Mendelian discovery.* Cassell, London.

(32) DARROW, K. 1941. Review in *Nature,* 147, p. 367, of "Science, philosophy and religion; a symposium", New York, 1941, including article by Darrow.

(33) DARWIN, C., assisted by F. DARWIN. 1880. *The power of movement in plants.* Murray, London.

(34) DARWIN, C. 1888. *Insectivorous plants*. Murray, London. (The first edition was published in 1875.)

(35) DARWIN, C. 1913. *The descent of man, and selection in relation to sex*, 2nd edition, reprinted. Murray, London.

(36) DARWIN, F. 1888. *The life and letters of Charles Darwin*. 3 vols. Murray, London.

(37) DARWIN, F. and A. C. SEWARD. 1903. *More letters of Charles Darwin: a record of his work in a series of hitherto unpublished letters*. 2 vols. Murray, London.

(38) DESOUTTER. 1941. Advertisement in *Daily Telegraph and Morning Post*, 11th June, 1941.

(38a) DISNEY, A. N., C. F. HILL and W. E. WATSON-BAKER. 1928. *Origin and development of the microscope*. Royal Microscopical Society, London.

(39) DRAPER, J. W., quoted by Andrade (6).

(40) DRIESCH, H. 1929. *Man and the Universe*. Allen and Unwin, London.

(41) EDWARDS, G. 1941. "Sydney Camm, creator of the Hawker Hurricane." *News Chronicle*, 18th February.

(42) EHRLICH, P. 1886. (No title.) *Zeit. wiss. Mikr.*, 3, p. 150.

(43) EINSTEIN, A. and L. INFELD. 1938. *The evolution of physics: the growth of ideas from the early concepts to relativity and quanta*. Cambridge University Press.

(44) ELTON, C. 1930. *Animal ecology and evolution*. Clarendon Press, Oxford.

(45) ELTON, C. 1933. *Exploring the animal world*. Allen and Unwin, London.

(46) FABRE, A. 1921. *The life of Jean Henri Fabre, the entomologist, 1823–1910*. Dodd, Mead, New York.

(47) FINDLAY, A. 1934. *The spirit of chemistry: an introduction to chemistry for students of the liberal arts*. Longmans, Green, London.

(48) FISHER, J. (No date.) *Birds as animals*. Heinemann, London.

(49) FORD, E. B. 1937. "Problems of heredity in the Lepidoptera." *Biol. Rev.*, 12, p. 461.

(50) FORD, E. B. 1938. *The study of heredity*. Butterworth, London.

(51) GELMO, P. 1908. "Über Sulfamide der p-Amidobenzolsulfonsäure." *Journ. prakt. Chem.*, 77, p. 369.

(51a) GLADSTONE, J. H. 1874. *Michael Faraday*. 3rd edition. Macmillan, London.

(52) GLASSER, O. 1933. *Wilhelm Conrad Röntgen and the early history of the Roentgen rays.* Bale, Sons and Danielsson, London.

(53) GODLEE, R. J. 1917. *Lord Lister.* Macmillan, London.

(54) GRANT, R. 1852. *History of physical astronomy, from the earliest ages to the middle of the nineteenth century.* Baldwin, London.

(55) GREGORY, R. A. 1916. *Discovery, or the spirit and service of science.* Macmillan, London.

(56) GREGORY, R. A. 1941. "Science and the structure of Society." *Nature*, 148, p. 4.

(57) GREGORY, R. (A.). (On behalf of the British Association.) 1941. "The commonwealth of science." *Nature*, 148, p. 393.

(58) HALDANE, J. B. S. 1932. *The inequality of man and other essays.* Chatto and Windus, London.

(59) HALDANE, J. B. S. 1938. *The Marxist philosophy and the sciences.* Allen and Unwin, London.

(60) HARRISSON, T. 1941. "Mass astrology." *New Statesman*, 16th August.

(61) HARRISSON, T. 1941. "Is science special?" *New Statesman*, 18th October, p. 364.

(62) HARTING, J. E. 1898. *The Rabbit.* Longmans, Green, London.

(63) HEATH, T. L. 1897. *The works of Archimedes.* Cambridge University Press.

(64) HENDERSON, G. C., A. J. GREENAWAY, J. F. THORPE and R. ROBINSON. 1932. *The life and work of Professor William Henry Perkin.* Chemical Society, London.

(65) HESSE, R. 1937. *Ecological animal geography.* Chapman and Hall, London.

(66) HILL, A. V. 1933. "International status and obligations of Science." *Nature*, 132, p. 952.

(67) HOBBES, T., quoted by Huxley, T. H. (79).

(68) HOGBEN, L. T. 1920. "Studies on synapsis. I. Oogenesis in the Hymenoptera." *Proc. Roy. Soc. B*, 91, p. 268.

(69) HOGBEN, L. T. 1920. "On certain nuclear phenomena in the oocytes of the gall-fly Neuroterus." *Trans. Linn. Soc. Zool.*, 34, p. 327.

(70) HOGBEN, L. T. 1921. "Studies on synapsis. III. The nuclear organization of the germ cells in Libellula depressa." *Proc. Roy. Soc. B*, 92, p. 60.

(71) HOGBEN, L. (T.) 1938. *Science for the citizen*. Allen and Unwin, London.

(72) HOLMES, S. J. (No date.) *Louis Pasteur*. Chapman and Hall, London.

(73) HOPKINS, F. G. 1935. "Science in modern life." *Nature*, 136, p. 893.

(74) HOWARD, H. E. 1920. *Territory in bird life*. Murray, London.

(75) HOWARD, H. E. 1929. *An introduction to the study of bird behaviour*. Cambridge University Press.

(76) HUMBOLDT, W. VON, quoted by Mill (98).

(77) HUXLEY, T. H. 1854. Review of "Vestiges of the natural history of creation. 10th edition. London, 1853." *Brit. For. Med.-Chir. Rev.* 13, p. 425.

(78) HUXLEY, T. H. 1855. "On certain zoological arguments commonly adduced in favour of the hypothesis of the progressive development of animal life in time." *Roy. Inst. Proc.*, 2, p. 82.

(79) HUXLEY, T. H. 1877. "On the study of biology." *Nature*, 15, p. 219.

(79*a*) ILTIS, H. 1932. *Life of Mendel*. Allen and Unwin, London.

(80) INFELD, L. 1941. *Quest: the evolution of a scientist*. Gollancz, London.

(81) JAFFE, B. 1934. *Crucibles: the lives and achievements of the great chemists*. Jarrolds, London.

(82) JEANS, J. 1937. *Science and music*. Cambridge University Press.

(83) JONES, R. A. 1941. "The man of science as aristocrat." *Nature*, 147, p. 677.

(83*a*) KOBBÉ, G. 1925. *The complete opera book*. Putnam, London.

(84) KROPOTKIN, PRINCE. Quoted by Marchant (92).

(85) LANKESTER, E. R. 1880. "On a new jelly-fish of the order Trachomedusae, living in fresh water." *Nature*, 22, p. 147.

(86) LANKESTER, E. R. 1880. "The freshwater Medusa." *Nature*, 22, p. 177.

(86*a*) LISTER, J. 1909. Article on "Joseph Jackson Lister" in *Dictionary of National Biography*. Smith, Elder, London.

(87) LITCHFIELD, H. 1915. *Emma Darwin: a century of family letters, 1792–1896*. 2 vols. Murray, London.

K

(88) LODGE, O.   Quoted by Andrade (6).

(89) LYND, R. S.   Quoted by Waddington (139).

(90) M——, H. G.   1941.   *Book review in Journ. Soc. Pres. Fauna Emp.*, 42, p. 52.

(91) MACLEOD, J. J. R.   1924.   "Insulin."   *Physiol. Rev.*, 4, p. 21.

(92) MARCHANT, J.   1916.   *Alfred Russel Wallace: letters and reminiscences.*   Cassell, London.

(93) MASSON, J. I. O.   1940.   "The universities and research." *Nature*, 145, p. 855.

(94) MCKIE, D.   1935.   *Antoine Lavoisier, the father of modern chemistry.*   Gollancz, London.

(95) MEES, C. E. K.   1929.   Article on "Photography" in the *British Encyclopaedia*, 14th edition.

(96) MIDDLETON, A. D.   1930.   "The ecology of the American grey squirrel (*Sciurus carolinensis* Gmelin) in the British Isles."   *Proc. Zool. Soc.*, p. 809.

(97) MIDDLETON, A. D.   1931.   *The grey squirrel.*   Sidgwick and Jackson, London.

(98) MILL, J. S.   1859.   *On liberty.*   Reprinted in *Utilitarianism, liberty and representative government* in Everyman's Library. Dent, London.

(99) MILL, J. S.   1863.   *Utilitarianism.*   Reprinted in *Utilitarianism, liberty and representative government* in Everyman's Library. Dent, London.

(100) MILLER, E. C.   1938.   *Plant physiology with reference to the green plant.*   McGraw Hill, London.

(101) MILLER, H.   1841.   *The old red sandstone.*   Reprinted in Everyman's Library. Dent, London.

(102) MILLER, H.   1881.   *Footprints of the creator.*   Nimmo, Edinburgh.

(103) MOORE-BRABAZON, J. T. C.   1941.   "The man of science as aristocrat."   *Nature*, 147, p. 544.

(104) MORGAN, T. H.   1907.   *Experimental zoology.*   Macmillan, London.

(105) MORTIMER, R.   1940.   "Books in general."   *New Statesman and Nation*, 2nd March, p. 277.

(106) MORTIMER, R.   1941.   "Books in general."   *New Statesman and Nation*, 18th October, p. 363.

(107) PARSONS, T. R. 1930.   *The materials of life.*   Routledge, London.

(108) PEARSON, K.   1900.   *The grammar of science.*   Black, London.

(109) PLUTARCH.   Quoted by Whymper, E. (144).

(110) POLANYI, M. 1939. *Rights and duties of science.* Manch. School econ. soc. stud., October, p. 175.

(111) POLANYI, M. 1940. *The contempt of freedom: the Russian experiment and after.* Watts, London.

(112) REMSEN, I. and C. FAHLBERG. 1879–80. "On the oxidation of substitution products of aromatic hydrocarbons." *Amer. Chem. Journ.* 1, p. 426.

(113) REYNOLDS, Q. 1941. Quoted in *News Chronicle.* 11th August.

(114) RIMSKY-KORSAKOFF, N. A. Quoted by Kobbé (83a).

(115) ROHR, M. VON 1936. *Abbe's apochromats.* Zeiss, Jena.

(116) ROMANES, G. J. 1880. "Physiology of the freshwater Medusa," *Nature,* 22, p. 179.

(117) ROMANES, G. J. 1882. *Animal intelligence.* Kegan Paul, Trench, London.

(118) RUSSELL, B. 1941. *Let the people think: a selection of essays.* Watts, London.

(119) S——, R. A. 1869–70. (Obituary notice of Graham.) *Proc. Roy. Soc.* 18, p. xvii.

(120) SCHULTZ, G. 1890. "Bericht über die Feier der Deutschen Chemischen Gesellschaft zu Ehren August Kekulé's." *Ber. deut. chem. Ges.* 23, p. 1265.

(121) SKENE, M. 1935. *A flower book for the pocket.* Oxford University Press.

(122) SPENCER, H. 1869. *The classification of the sciences, to which are added reasons for dissenting from the philosophy of M. Comte.* Williams and Norgate, London.

(123) STURGES, T. W. 1915. *The poultry manual: a complete guide for the breeder and exhibitor.* Macdonald and Evans, London.

(124) SWIFT, J. 1726. *Travels into several remote nations of the world by Lemuel Gulliver, first a surgeon, and then a captain of several ships.* Reprinted by Penguin Books, Harmondsworth, 1938.

(125) SWINTON, W. E. 1934. *The Dinosaurs, a short history of a great group of extinct reptiles.* Murby, London.

(126) TANSLEY, A. G. 1929. *The new psychology and its relation to life.* 11th impression. Allen and Unwin, London.

(127) TAYLOR, F. S. 1940. *The conquest of bacteria: from 606 to 693.* Secker and Warburg, London.

(128) THIMANN, K. V. and J. B. KOEPFLI. 1935. "Identity of the growth-promoting and root-forming substances of plants." *Nature,* 135, p. 101.

(129) THORNDIKE, L. 1923. *A history of experimental science.* Vol. 1. Macmillan, New York.

(130) THORPE, E. 1911. *Essays in historical chemistry.* Macmillan, London.

(131) THORPE, E. 1926. *A dictionary of applied chemistry.* Vol. 6. Longmans, Green, London.

(132) TILDEN, W. Quoted by Gregory (55).

(133) TILDEN, W. A. 1922. *Chemical discovery and invention in the twentieth century.* 4th edition. Routledge, London.

(134) TRÉFOUËL, J. and J., F. NITTI and D. BOVET. 1935. "Activité du *p*-aminophénylsulfamide sur les infections streptococciques experimentales de la souris et du lapin." *Compt. rend. Soc. biol.* 120, p. 756.

(135) TRUEMAN, A. E. 1938. *The scenery of England and Wales.* Gollancz, London.

(136) VALLERY-RADOT, R. 1900. *La vie de Pasteur.* Hachette, Paris.

(137) VERULAM, FRANCIS (LORD). 1658. *Sylva Sylvarum: or, A Natural History.* 7th edition, edited by W. Rawley. Lee, London.

(138) VERULAM, FRANCIS, LORD. (No date, probably 1660.) *New Atlantis. A work unfinished.* Edited by W. Rawley. Lee, London.

(139) WADDINGTON, C. H. 1941. *The scientific attitude.* Harmondsworth (Pelican).

(140) WALLACE, A. R. 1894. *The Malay archipelago.* Macmillan, London.

(141) WATSON-BAKER, W. E. 1941. *Personal communication.*

(141a) WELLS, H. G. 1941. "The man of science as aristocrat." *Nature,* 147, p. 465.

(142) WENT, F. W. and K. V. THIMANN. 1937. *Phytohormones.* Macmillan, New York.

(143) WEST, E. S. 1941. "Beethoven and the future." *New Statesman and Nation,* 5th April, p. 363.

(144) WHYMPER, E. (No date.) *Scrambles amongst the Alps in the years 1860–69."* 5th edition. Nelson, London.

(145) ZAREK, O. 1941. *German odyssey.* Cape, London.

(146) ZOETHOUT, W. D. and W. W. TUTTLE. 1940. *A textbook of Physiology.* 7th edition. Kimpton, London.

Index

THE SCIENTIFIC LIFE

SCIENCE AND THE
PLANNED   STATE

SCIENCE

and the

PLANNED STATE

by

JOHN R. BAKER

M.A., D.Phil., D.Sc.

NEW YORK

THE MACMILLAN COMPANY

1945

PRINTED IN THE UNITED STATES OF AMERICA

Preface

In a sense this book is a sequel to *The Scientific Life*, which was published in 1942. Care has been taken to make it equally suitable for those who have read the earlier volume and those who have not. There is very little repetition, though I have once more insisted on the importance of chance in scientific discovery. The main purpose of *The Scientific Life* was to describe the human nature of the good research worker, and to show that it would be futile to try to confine him within the rigid boundaries of a central plan for the advancement of science. Both books are meant for scientist and layman alike, including the older students at schools, and both are lightly written; but this new book is much the more systematic in treatment. It consists of a careful analysis and criticism of the totalitarian view of science, of which so much has been heard in recent years.

It is a great pleasure to acknowledge the help given me by my wife and my friends, especially Dr C. P. Blacker, M.C., F.R.C.P., Professor V. H. Blackman, F.R.S., Professor E. S. Goodrich, F.R.S., Professor M. Polanyi, F.R.S., and Professor A. G. Tansley, F.R.S. Professor Blackman has read the whole book and the others have read parts of it, and all have made valuable suggestions.* Even more valuable has been their encouragement, for the book is in conflict with a social outlook that has much popular support to-day. The person who wants to be comfortably reassured in the beliefs current in our newspapers will find little to his liking in these pages. I address myself to the boy or girl, man or woman who has the courage and independence to think for himself or herself and to question the validity of popular opinion.

* It is not implied that all the friends mentioned agree with every statement in the parts of the book they have read.

9

Chapter II is an expansion of an address delivered to International Student Service, and parts of Chapters IV and V contain material presented in a lecture to the Institute of Physics. I thank I.S.S. and the Institute for permission to publish this version of what I said.

JOHN R. BAKER.

BURNT OAK,
  KIDLINGTON,
    NR OXFORD.

Contents

Chapter I

INTRODUCTION

1. *The Story of John Ellis, Merchant and Naturalist*

JOHN ELLIS, London merchant, had a most unusual hobby. He received seaweeds from collectors on the Welsh and Irish coasts. "These, when properly dried, I disposed on thin Boards covered with clean white Paper, in such a manner as to form a kind of Landscape, making use of two or three Sorts of the *Ulva marina*, or Sea-Liverwort of different Colours, in designing a Variety of Hills, Dales, and Rocks, which made a proper Ground-work and Keeping for the little Trees, which the expanded Sea-plants and Corallines not unaptly represented." [27]

A friend expressed pleasure in viewing these landscapes and encouraged him to make some for the royal family, so that with the aid of examples the young princesses might try the new art for themselves. The landscapes were accepted with royal condescension, though there is no evidence that anyone has ever copied them during the space of nearly two centuries which separates us from these events.

To supply the princesses in a fashion that would accord with their exalted rank, Ellis undertook a large collection of marine plant and plant-like productions. He soon began to want to classify them in a natural way. He turned at first to the system of the illustrious Ray, but discovered that the latter had erred in attributing them all to the vegetable kingdom. Ellis found it necessary to use a simple microscope, and it was not long before he discovered that in many of them the "Texture was such, as seemed to indicate their being more of an animal, than vegetable Nature." Soon he "was convinced, from my own Observations of the Subjects themselves, that several, which had hitherto been considered by Naturalists, as Marine Vegetables, were in Reality of animal Production."

13

Ellis was now like a hound on hot scent. He undertook visits to the coasts, accompanied by artists who drew what they saw through "a very commodious Microscope." Very remarkable were the organisms that his commodious instrument presented to their view, and remarkably accurate and beautiful their drawings. Without in the least minimizing the previous studies of Jean-André Peyssonnel in the Mediterranean, of Abraham Trembley in Holland, or of Bernard de Jussieu on the coasts of France, it may justly be claimed that Ellis, more than any other single man, caused scientists to accept into the animal kingdom the great and diverse groups of hydroids, sea-fans and their allies, and Polyzoa. No other man has ever made so great an enlargement of the subject-matter of zoology.

His contemporaries were not slow to acknowledge the services to science of this gifted amateur, whose investigations had had such an unpredictable beginning. His *Essay towards a Natural History of the Corallines* [27] was published in 1755, and by 1768 he was not only a Fellow of the Royal Society, but had received from its President the Copley Medal and an enthusiastic address of congratulation. These encouragements made Ellis even "more anxious in the pursuit of his favourite study."

Like many investigators of nature, Ellis pursued science as an end in itself and was also interested in its applications. His best work was purely scientific, but he also did much to apply botanical knowledge to practical human affairs. He published accounts of the mangosteen, breadfruit, and other plants, with directions for conveying seeds and seedlings to distant parts of the world to promote the purposes of medicine, agriculture, and commerce. His historical account of coffee, published in 1774, was intended to increase the consumption of that article and thus help the planters in the West Indies. So it was written of him, "And his active mind was constantly employed in devising means for promoting the welfare of society until the time of his death, which happened on the 15th of October 1776."[28]

These activities had their influence on contemporary society, but Ellis is remembered to-day almost entirely for his purely scientific work. He was intensely interested in plant-like animals, and he wanted to share his interest with others. That is how he regarded the social aspect of science, apart from its applications. He said so in no uncertain terms. "And now," he wrote, "should it be asked, granting all this [about plant-like animals] to be true, to what end has so much labour been bestowed in this demonstration? I can only answer, that as to me these disquisitions have opened new scenes of wonder and astonishment, in contemplating how variously, how extensively life is distributed through the universe of things: so it is possible, that the facts here related, and these instances of nature animated in a part hitherto unsuspected, may excite the like pleasing ideas in others." [28]

2. *Definitions*

I have started by relating the story of John Ellis because it is in such sharp conflict with the totalitarian view of science and thus makes our subject vivid. The story brings home a fact about science which could be illustrated less strikingly but equally truly from the personal experience of many investigators, great and small—the fact that scientific discovery is often prompted by extraordinary and unpredictable circumstances apart from the material needs of man. These needs are served by science, but the promptings of many investigators are neither economic nor materialistic. Szent-Györgyi, who was awarded the Nobel Prize for Medicine, has made some forcible remarks bearing directly on this subject. "The results of my own scientific activity," he says, "have restored many men to life and health and preserved others from illness, and therefore I have often been praised for my goodness. When I hear this I cannot help smiling." [98] He goes on to say that "the man who sets out to work intending to discover something useful" should actually be excluded

from the laboratory.   He remarks that every real discoverer
"was spurred on by an inner passion and impulse."

We are faced to-day by a considerable body of propaganda
intended to make us view science in a light which Ellis would
not have understood.   We are told that the science of any
given period is the inevitable result of political and economic
conditions.   The scientist may think that he is guided by his
interests, but actually his studies are determined by material
forces beyond his control.   This theory forms part of the
materialist view of science, accepted by the rulers of totalitarian
states.   It is the purpose of this book to examine the materialist
or totalitarian view.

It is necessary at the outset to be sure about what will be
meant by certain constantly recurring words.

It is sometimes said that there is no such thing as pure science.
This is an argument about the meanings of words, not about
facts or opinions.   On the one hand there is an organized body
of general demonstrable knowledge about the matter of the
universe.   Students of this body of knowledge unhesitatingly
recognize a hierarchy of importance among its parts, according
to the amount of light they throw on other parts and the degree
in which they are capable of being comprehended under
generalizations.   On the other hand there is a second body of
knowledge which codifies the applications of part of the first
body of knowledge to material human welfare and also includes
many demonstrable facts that are used to supply the material
wants of man but not to throw light on other facts nor to
support generalizations.   The distinction between the two
bodies of knowledge is so necessary and obvious that even
those who deny it nevertheless find it necessary to draw it in
their writings.   They may try to get over the difficulty by
writing "pure" science, with pure in inverted commas, or by
various circumlocutions, but they know that a distinction
exists.

There is no reason why a generalization, capable of linking
facts not previously known to be related, should be materially

beneficial to man. Similarly, to say that every demonstrable discovery will be used in a practical way before the inevitable extinction of mankind is to make a prophecy that is without any foundation. Again, there are technical processes of material benefit to man which are not understood, and have not thrown light on other facts. Two bodies of knowledge exist, which differ although they overlap. In the one, the best is what is most enlightening: in the other, the best is what works best.

If it be wished to avoid all purely linguistic arguments, the first body of knowledge may be called *a* and the second *b*. Anyone who wishes to group the two under a single title is at liberty to say that $a + b = c$. Words, however, are better than symbols, because more easily remembered, and *science* is a convenient word for *a* and *technology* for *b*. It is an unfortunate fact that the convenience of words as compared with symbols makes people quarrel over words when they think they are quarrelling over facts or opinions about facts. Thus, when *a* is called science, it is argued that it is presumptuous to confine the meaning of this word to only a part of *c*. Again, when *a* is called pure science, it is queried whether *b* is impure or tainted. This query was raised repeatedly by different speakers at the meeting of the British Association held at the Royal Institution, London, in September 1941. Such talk is not argument, but only a ridiculous kind of playing with words. In this book, *science* will be used for *a* and *technology* for *b* purely as labels, not as arguments about anyone's opinion as to the value of the thing named. Technology is the knowledge of techniques which serve man's material wants. It consists partly of the applications of science, and partly of empirical and rule-of-thumb knowledge about isolated facts.

While we are on the subject of words, it may be remarked that the person who cultivates science is a *scientist*. A curious opposition has arisen to the use of this convenient word, and multisyllabic substitutes are sometimes preferred. People who would blush to use the expressions *men of music* or *musical men* do not hesitate to write *men of science* or *scientific men* when they mean scientists.   This post-prandial pomposity is best relegated to the environment of the white tie and tail-coat.   It is true that the expression "Men of Scyence" is to be found in a sixteenth-century writing, but the context suggests that it was then applied to technologists.   The term *scientific workers*, again, is three syllables too long, and although not so obnoxious as the others, yet lends itself to use by those who would surreptitiously insinuate some special bond of unity between scientist and factory-worker.

Though I have lived a quarter of a century in scientific circles, and seen men of science and scientific men and scientific workers tumbling over one another in print, I do not recollect ever having heard any of these expressions used in conversation (except when *The Scientific Worker* was used as the name of a journal).   The word always or almost always used in con-versation is *scientist*.   We owe that word to the Rev. William Whewell, Professor of Moral Philosophy in the University of Cambridge.   Few deliberately invented words have gained such wide currency, and many people will probably be sur-prised to learn that it is only just over a century old.   Whewell undertook the invention of the word in no light spirit.   His "Aphorisms concerning the Language of Science" cover more than seventy pages of *The Philosophy of the Inductive Sciences.*[109] The "Aphorisms" constitute a careful inquiry into the ways in which new words are and should be constructed.   He observes that the terminations *ize*, *ism*, and *ist* are applied to words of all origins.   Forthwith he coins the universally accepted *physicist*, remarking that *physician* cannot be used in this sense.   He proceeds at once to the invention of an even more necessary word.   "We need very much a name to describe a cultivator of science in general.*   I should incline to call him a *Scientist*."   So should I.

So much for the words *science* and *scientist*: now for *totalita-*

* The context makes it clear that Whewell did not restrict the name scientist to those whose interests cover the whole subject.

rianism. This ugly mouthful has unfortunately no euphonious synonym. The word had gained general currency as a comprehensive name for the political systems of Germany, Italy, and the U.S.S.R., when Hitler's invasion of Russia made the astute editors of our daily and weekly Press recognize that it might be inept to bracket our new allies with our old enemies. The word was dropped like a hot brick. It has cooled too long and I pick it up without hesitation. By totalitarianism I mean those systems of government in which the actions of individuals are to a great extent controlled by a central planning authority. It is the antithesis of anarchy, but as that is a system which no country has ever adopted, the most exact opposite in the world of reality is liberalism. It is strange to reflect that there is so little memory to-day of what liberalism stands for, that the man in the street thinks of it as intermediate between socialism and conservatism. This idea is misleading. If we wish to arrange the various political systems in linear order, it may be suggested that liberalism should be placed beyond, not between, socialism and conservatism. At one pole come the totalitarian systems (nazism, fascism, and communism), in which the state is all-powerful and ruthless and the individual deprived of liberty. Next come socialism (in the narrower sense) and conservatism, under both of which the state has great power but avoids ruthlessness and allows some liberties to individuals. At the opposite pole to totalitarianism stands liberalism, the system which puts the liberty of the individual above all else and regards the state merely as a mechanism for minimizing people's interference with one another's freedom. It is in this sense, then, of antithesis to individual liberty that the word totalitarianism is used in this book.

3. *The Case for Totalitarianism in Science*

The scientist, like all other members of the community, has rights and responsibilities. The rights are the conditions which the community must grant to him if he is to be able to perform his function effectively. If he is a teacher, these rights resemble those of all teachers and do not call for special comment here. If he is a research worker, they call for a lot of comment. The history of science and the experience of modern research suggest that the investigator's rights, above everything else, are four kinds of freedom: freedom to become a research worker, freedom of association, freedom of inquiry, and freedom of speech and publication. Most of these freedoms have been lost in totalitarian countries, and the ever-growing movement towards central planning threatens them in Britain.

Balancing the scientist's rights are his responsibilities. There are, of course, obvious duties, which differ according to the nature of the function which the scientist is fulfilling. If he is a teacher of science at school or university, a primary duty is clearly to teach science to the best of his ability and imbue his pupils with its spirit. If he is a research worker, his duty is naturally to make the greatest possible contribution to demonstrable knowledge. Many members of the staffs of universities are both teachers and research workers and thus have both these duties. There exists also a wide field of further opportunity to repay the community for its grant of the freedoms that enable the scientist to live a full life and exploit his talent. These opportunities are open to all scientists, whether students or teachers or research workers.

A lot of trouble has been taken to tell scientists what their responsibilities are. Large volumes have been written and addresses given to considerable congregations. Nevertheless, it is not easy to find a formal statement of the obligations of the scientist to the community, as laid down by our modern scientific moralists. It is therefore necessary to try to condense the message into a form in which it can be examined. It seems to be divisible into three propositions:

(1) Science exists to serve the material wants of man.

(2) Central planning makes for efficiency, and scientists have a duty to press for its introduction in their own sphere. Not only would overlapping be avoided, but trivial investigations would no longer take up time and money. The central authority would know the material wants of the community, and would be able to direct the course of research in such a way as to relieve these wants as speedily and economically as possible. Teams of research workers would be available to be switched quickly on to the most pressing need of the day.

(3) Since scientists are accustomed to do everything methodically, and since they must recognize that central planning would improve the efficiency of their own work, they should press for the adoption of a scientific scheme of central planning in all departments of life. "Logically," says Bukharin, "Marxism is a scientific system, a scientific outlook and scientific practice."[4] Scientists have been irresponsible, supporting any political party or none, as though they had no serious concern with the welfare of others. It is their duty to take an interest in party politics and to support the side that promises to introduce a policy of central planning under state control. It will also be to their own advantage to ally themselves with the party that will gain power.

The purpose of this book is to consider these three propositions. When they are stated like this, in moderate terms, a case is presented which calls for careful consideration. Those who have brought the arguments forward in moderate form have performed a useful service by making scientists reflect seriously about their social obligations. Nevertheless, the arguments are sometimes stated in an immoderate form which is not only not helpful, but positively threatening. It is important that scientists should understand clearly how extreme are the views of some of those who profess to teach ethical principles to scientists. These views are available in print, but are not generally known. It is desirable that scientists who are influenced by arguments about their morals should understand clearly what ideas influence the extreme wing of their ethical authority. They can find out easily enough by referring to Mr J. G. Crowther's book *The Social Relations of Science*, published in 1941 by Messrs Macmillan.[19]

Mr Crowther is of opinion that scientists should attach themselves to certain particular political groups. The use of persuasive argument is not thought likely to suffice, for Mr Crowther says quite unequivocally that under some circumstances the use of inquisition is desirable. "Inquisition is beneficial to science," he says (p. 333), "when it protects a rising class." The reader should refer to the book itself, so that he may not be under any misapprehension as to what Mr Crowther means. He means that the methods of the Inquisition of ancient times should be reintroduced to make sure that people shall accept political views in accord with his own. "Those who have revived the Inquisition, like the Pope in Galileo's time," he writes (p. 331), "have a better understanding of politics [than most scientists of to-day], and realize that in crises the possession of power is more important than the cultivation of intellectual freedom. . . . The danger and value of an Inquisition depend on whether it is used in behalf of a reactionary or a progressive governing class." A "progressive governing class," in fact, is right in using the methods of the Inquisition, and scientists would be well advised to get on the safe side in politics. Mr Crowther does not find fault with authority for its treatment of Galileo, nor with that great scientist for recanting when threatened with torture, but only blames Galileo (p. 333) for not "seeking the protection of progressive powers, who will fight for him as well as argue, if necessary."

The reader may be inclined to minimize the influence of Mr J. G. Crowther in the scientific world. That would be a mistake. He is the secretary of the scientific section of the British Council, and thus in charge of the representation of British science before the world. The extreme section of opinion is unquestionably influential.

Nothing less than a realization of the gravity of the situation would induce me to copy those with whom I disagree by ascending the pulpit. Some one, however, has got to do it. Opinion is inevitably formed if all the talking and writing come from one side. Only one point of view is being presented as to the moral obligations of scientists. That point of view leads by insensible steps to the conclusion that scientists should be subjected to inquisition to secure conformity with political dogmas. By a twisting of the English language a policy that involves reversion to the cruelties of the Middle Ages is presented as the course of progress.

The purpose of this book is to suggest that progress lies in another direction. The design is as follows. Chapter II will be devoted to a consideration of the argument that science exists only to serve the material wants of man, upon which the whole of the totalitarian view of science is based. The freedoms which serve best the cause of discovery will be discussed in Chapter III. Chapter IV will provide an actual example of what happens to science when these freedoms are denied. So much for the rights of scientists: the last Chapter will be concerned with their duties and responsibilities, and a general survey will be given of the ways in which they may use their special talents for the benefit of others.

Chapter II

THE VALUES OF SCIENCE

"Other interests, besides the material wants of life, occupy the minds of men."—A. VON HUMBOLDT.[50]

1. *Grades of Opinion on the Values of Science*

THAT scientific knowledge can be applied to the material welfare of man is so obvious that no discussion of this value of science is necessary. Those who think that science has other values do not minimize its contributions to the feeding of human beings and their protection from the elements and from ill-health. There are those, however, who deny that science has any value apart from these contributions to material welfare. Four grades may be distinguished in the scale of opinion, as follows:

(1) Science has value only in serving the material wants of man. The only consideration is the material welfare of the community as a whole. This is the extreme totalitarian position.

(2) Science has value only in serving the material wants of man, but research workers do their best work if they enjoy it.

(3) Science has value both in serving the material wants of man, and also in enabling people to escape from certain mental evils. The study of science prevents unhappiness consequent on pettiness of outlook, and produces forgetfulness of unpleasant memories. This rather negative position is that of Bertrand Russell.[91]

(4) Science has another value besides serving the material wants of man and enabling people to escape certain mental evils. It has a positive primary value as an end in itself, like music, art, and literature.

In the past there also existed a fifth opinion, held by those who were actually glad when they thought that .certain scientific discoveries could not be used to serve the material wants of man. This opinion scarcely exists to-day and seems not to merit further consideration.

2. *Scientists do not work only for Material Ends*

There is not any necessary connexion between the material usefulness and intrinsic interest of a scientific discovery. "We can declare without the least hesitation," says Szent-Györgyi, the famous biochemist, "that to judge scientific research by its usefulness is simply to kill it. Science aims at knowledge, not utility." [98] It is extremely unlikely that every discovery will serve man in a material way before the inevitable extinction of human life. Some of the most profound truths will probably not be used practically. Professor G. H. Hardy [43] has made this point neatly for mathematics. He cites some easily understood mathematical proofs, whose beauty and general significance are apparent to everyone who follows them. Having won the reader's willing assent to their value, he goes on to prove that they not only are not, but cannot be used by the practical man. Euclid's proof that the number of prime numbers is infinite is so masterly and. economical that everyone who follows it, mathematician or not, acclaims its value; but as Professor Hardy points out, it is more than sufficient for the engineer to know that the number of primes less than 1,000 million is 50,847,478, for practical men never work to more significant figures than this. In science we can never say that a discovery will never be used to promote material welfare, but we can and must say that scientists are interested in discoveries apart from the possibility of their producing food, shelter, health, etc.

The pretence that science only serves humanity by giving us food, health, and shelter leads to nonsense; for it means that we live only for food, health, and shelter, instead of requiring them so as to live for something else. Why do we feed and protect ourselves and others? Is it so that we and they may live to feed and protect others, so that they may do the same for yet others, and so on interminably and senselessly? "Have we nothing eventually in view more admirable than the abolition of want and the securing of comfort for everyone, ends which at present bulk so large in our programs?" The question is put by the distinguished American physicist, Professor P. W. Bridgman. "Will we be permanently satisfied with these, or will something more be necessary to give dignity and worth to human activity?" [12]

There must be something else for which people want to live. Great music, art, and imaginative literature, it may be suggested, are examples of valid ends. If a scientist makes that answer, it is necessary for him to say that he practises science so that the applications of what he discovers may keep people alive, so that they may appreciate music, art, and literature, which are the real ends in life which make him practise science. This house-that-Jack-built rigmarole is nonsense. The scientist may indeed value these subjects highly, and they are certainly ends in themselves; but if his dominant impulse were not scientific he would be a poor scientist. Science is as much an end in itself as music or art or literature. ". . . if ever there are ends in themselves or goods in themselves," Professor Bridgman has written, "then surely the gratification of the craving for understanding is one of them." [12]

People engaged in practical pursuits have often advanced science, and this fact is sometimes made the basis of a claim that science had its origin in a desire to satisfy the material wants of ordinary human life. From that premiss it is argued that scientists should devote themselves to the satisfaction of those wants. Even if the claim were justified, the conclusion could not be logically deduced from it; but the claim itself is not justifiable. We cannot know anything for certain about the earliest beginnings of science, but we do know that modern savages are interested in natural objects and phenomena apart from their material usefulness. Science as we know it to-day may be said to have originated about the eighteenth century, for although there were scientific geniuses before then, the spirit of the subject was confined to a small number of people, and their discoveries were somewhat isolated. During that century there was a wonderful blossoming forth of science. Magnificent work was done, especially in biology. The best of that work was inspired by nothing but an intense desire for knowledge for its own sake.

The scientist of to-day is often cynically indifferent to the early history of his subject. He knows that people used to make fantastic concoctions intended to cure human ailments, and he recognizes no connexion between such activities and his own. He is right, but he has missed the point. The men who were struggling solely to give practical help to mankind often made little or no contribution to knowledge; but those who had an intense desire for knowledge for its own sake were doing research that is comparable with the very best that is being done to-day.

Just over two centuries ago Réaumur [88] published a memoir on the reproduction of aphids and Trembley [102] a book on the natural history and response to experimental procedures of the little fresh-water polyp, *Hydra*. I challenge anyone who is cynical about old-time science to point out any modern work that provides a better example of scientific method than those studies of Réaumur and Trembley. Réaumur's memoir was devoted to the question whether aphids can reproduce without sexual union. The way in which he tackled this question, in free collaboration with Bonnet, Bazin, Trembley, and Lyonet, provides an example to be copied by modern scientists. The clear introductory statement of what was already surmised on the subject, the scrupulous care and accuracy of the work, the elaborate attention to detail, the unwillingness to accept anything without stringent proof, the avoidance of unnecessary hypotheses—all these are models for all time. Réaumur and his friends established beyond question that aphids can reproduce without sexual union.   Trembley's book on *Hydra* is, of course, a classic.   It contains not only an excellent description of the form and natural history of the various species, but also a full account of the studies on regeneration, which may be said to mark the origin of experimental zoology.   Indeed, these experiments are quoted in modern text-books, not as historical curiosities, but as our best information on the subject.   Trembley's description of how he turned the minute organism inside out and how it survived this extraordinary operation was for long disbelieved; but recently Mr R. L. Roudabush [89] has succeeded in repeating Trembley's experiment and confirmed the survival of the reversed animals.   The whole of Trembley's book, like Réaumur's memoir, is a model of scientific method. *In neither is there any indication that the author was striving to satisfy the material wants of man.*   Their spirit was that which has been the chief animating influence of science ever since.

The scientist of to-day often opens a text-book and takes what he reads there as though it had arrived on those pages as a matter of course.   What an eye-opener it would be if he could glimpse, even vaguely, the history of the knowledge contained in a single sentence chosen at random!   Even if the sentence dealt with a modern subject, its history would go far back along the ages; and he would see a succession of the men who brought the knowledge contained in it into being.   They were not just names in a history-book of science: they were real, live people, diverse in many ways, but nearly all united in belief in the value of science as an end in itself.

3. *The Borderland between the Material and Immaterial*
Values of Science

There are certain values of science which stand half-way between the crudely material values on the one hand and the immaterial values on the other.   Knowledge of the facts of equal inheritance from both parents (apart from the genes borne on the sex-chromosomes) is important in framing people's general social outlook, but it does not directly provide them with food, health, or shelter. The relative status of man and woman would be different from what it is if people believed that inheritance were wholly maternal (as the Trobriand Islanders, for instance, are said to believe) or wholly paternal (as some biologists once thought). Again, people's social outlook is affected by their beliefs on the scientific question whether what are popularly called "acquired characters" are inherited. The function of science in reducing superstition comes into this category of values.

4. *The Appreciation of Science as an End in Itself*

We must now analyse the immaterial or spiritual values of science.

The history of science suggests that many great investigators have accepted the value of science as an end in itself as something so obvious as not to require analysis. Einstein has well expressed what are probably the inarticulate feelings of many people who value science as an end. "The satisfaction of physical needs," he writes, "is indeed the indispensable precondition of a satisfactory existence, but in itself it is not enough. In order to be content, men must also have the possibility of developing their intellectual and artistic powers to whatever extent accords with their personal characteristics and abilities."

There are reasons for thinking that science is potentially the greatest achievement of the human mind. Optimists may look for that greatest achievement in ethical perfection. They may be right and I hope they are; but life among savages has shown me that if civilization and religion have improved men morally, then the improvement that has occurred has been too small to give reason for much optimism about the future. In most intellectual fields we cannot look forward with confidence of progress. There is no reason to suppose that the historians of the future will tower above those of the present day. Philosophy has given the world some of its greatest geniuses, but the history of the subject contradicts the idea of a gradual approximation towards a consensus of opinion on philosophical subjects. We cannot guess the future of music, but at least it may be said that the world to-day has no composer who will bear comparison with the geniuses of the past. It is sometimes argued that geniuses are not recognized in their own times, and that we may even now have a genius of musical composition in our midst; but the fallacy of this argument is apparent to anyone who is acquainted with the history of music. The same considerations apply to pictorial art, and there is no sure ground for thinking that we are merely experiencing a phase of relative inactivity which will be followed by a new outburst of progress. In science, on the contrary, the present state of affairs and the prospect for the future are both very good. The standard of excellence is as high as ever it was. We have genius to rank with the greatest of all time (in physics alone we have Bohr, Dirac, Einstein and Schrödinger, and have only recently lost Rutherford and J. J. Thomson). If science be left free to expand, its expansion is inevitable, for science grows by accretion.

The unimportant composer or artist does nothing permanent to make his subject greater. The unimportant scientific research worker, on the contrary, places his brick firmly in position, and on it every subsequent worker in the same field —geniuses included—will build again. The knowledge that every step forward is an advance in a gigantic undertaking is an inspiration to the scientist, for he may legitimately feel that he is playing his part in the greatest adventure of the human mind. This knowledge is one of the supreme values of science to the investigator.

It is impossible to read the biographies of the greatest scientists without realizing the high value which they have attributed to science apart from its material benefits, but they seldom analyse their appreciation very explicitly. It is unquestionable that a pleasurable excitement in approaching the unfamiliar is a part of the reason for their appreciation, an attitude of mind which is shared with the geographical explorer. A pleasure in finding order where previously disorder seemed to reign is another component of the scientific attitude. This has been stated quite unequivocally by the Danish genius of physics, Niels Bohr,[11] who writes that the deepest foundation of science is "the abiding impulse in every human being to seek order and harmony behind the manifold and the changing in the existing world." T. H. Huxley wrote in his *Method and Results* that the research worker is inspired by "the supreme delight of extending the realm of law and order ever farther towards the unattainable goals of the infinitely great and the infinitely small, between which our little race of life is run."[53] Some scientists, again, are animated by a component of that special awareness of the natural environment and feeling of community with nature and joy in natural beauty which also animate the poet and artist in their respective fields. This was clearly understood by the great German scientist, Alexander Humboldt,[50] who wrote of "that important stage of our communion with the external world, when the enjoyment arising from a knowledge of the laws, and the mutual connexion of phenomena, associates itself with the charm of a simple contemplation of nature."

Humboldt was a person of extraordinarily wide interests. As a young man he was a successful mining technologist, but his passion for travel drew him into wider and wider fields of study until it might be said of him that if ever there was such a person as a general scientist, it was he. Few men, if any, have ever made such substantial contributions to so many diverse branches of science; and it was not only science that engaged his attention, for he was also a diplomat of high rank and a political economist. The extraordinary breadth of outlook of this great man enabled him to see science as a whole, and he expressed very vividly what he saw. In thirteen words of the utmost simplicity he expressed a truth which our modern materialists cannot shake: "other interests," he wrote,

"besides the material wants of life, occupy the minds of men."
He instanced the "desire of embellishing life by augmenting
the mass of ideas, and by multiplying means for their generalization. . . . The higher enjoyments yielded by the study of
nature depend upon the correctness and the depth of our
views, and upon the extent of the subjects that may be comprehended in a single glance." These words are strikingly
similar to those written by the philosopher, Alexander, not
much less than a century later: "The greatest truths are
perhaps those which being simple in themselves illuminate
a large and complex body of knowledge." Such truths, when
grasped, unquestionably bring pleasure to the mind; and it
would be fantastic to deny the existence of this kind of pleasure
or to assess it lower than crude or material kinds. "In considering the study of physical phenomena," said Humboldt,
"we find its noblest and most important result to be a knowledge of the chain of connexion, by which all natural forces are
linked together, and made mutually dependent upon each
other; and it is the perception of these relations that exalts
our views and ennobles our enjoyments."

The enjoyments appear subjectively to be of the same kind
as those caused by the perception of artistic beauty, combined
with wonder or even a pleasurable astonishment. Professor
J. B. S. Haldane [42] has stressed the value of beauty in science
in a particularly concrete way. "As a result of Faraday's
work," he wrote, "you are able to listen to the wireless. But
more than that, as a result of Faraday's work scientifically
educated men and women have an altogether richer view of
the world: for them, apparently empty space is full of the most
intricate and beautiful patterns. So Faraday gave the world
not only fresh wealth but fresh beauty." These simple words
express a profound truth, which can be denied only as a tone-
deaf man can deny the spiritual value of music. They are a
distinguished investigator's flat contradiction of the materialist
concept of science. Darwin expresses his feelings of beauty
and wonder in the final words of *The Origin of Species*: "There

is grandeur in this view of life, with its several powers, having been originally breathed by the Creator into a few forms or into one; and that, whilst this planet has gone cycling on according to the fixed law of gravity, from so simple a beginning endless forms most beautiful and most wonderful have been, and are being evolved." [22]

The finding of a kind of wonder or awe in the majesty and apparently infinite complexity of the universe has led some of the greatest scientists—among them Boyle, Hooke, Newton, and Trembley—to ascribe the value of science to its giving us an insight into the mind of God. The great Swiss-born American zoologist, Louis Agassiz, for instance, expressed this idea unequivocally: "If I mistake not, the great object of our museums should be to exhibit the whole animal kingdom as a manifestation of the Supreme Intellect." [47] This seems to be related to the subtler feeling of some of the greatest mathematicians that mathematical reality lies outside human beings, and that in their apparently creative work they are actually only observing and recording. [43]

The scientist is able to construct a sort of scale of scientific values and to decide that one thing or theory is relatively trivial and another relatively important, quite apart from any question of practical applications. There is, as Poincaré has well said, "une hiérarchie des faits." [81] Most scientists will agree that certain discoveries or propositions are more important because more widely significant than others, though around any particular level on the scale of values there may be disagreement. Thus, every scientist will agree that the discovery of atoms and of cells was important, and that the discovery of a new species of beetle, not markedly unusual in any way, is unimportant. So also with theories and "laws." A law, says Poincaré, "sera d'autant plus précieuse qu'elle sera plus générale." Professor G. H. Hardy [43] has shown how mathematicians value their ideas by generality and depth, and how they universally value general and deep theorems above mere isolated curiosities, such as the fact that 8712 and 9801

c

are the only four-figure numbers that are integral multiples of themselves written backwards ($8712 = 2178 \times 4$ and $9801 = 1089 \times 9$). A general theorem is one of wide significance, and a deep theorem one requiring a first understanding of a simpler theorem. Both these ideas are continually being used, consciously or unconsciously, whenever one scientist says that another has done a "good" bit of work.

The existence of amateur scientists is a proof that science is appreciated as an end. The amateur plays a smaller part in scientific research than he did in the eighteenth and nineteenth centuries, but excellent work is still done by amateurs in geology and biology (see pp. 90–91). Apart from those who are sufficiently interested to rank as amateur scientists, there is a mass of people who possess the same sort of feelings as the great investigator but in lesser degree. For instance, a markedly strange animal of any kind arouses great public interest in both savage and civilized communities, and no sharp dividing line can be drawn between this sort of interest and that which inspires the zoologist, though the latter's interest is of course greater and more lasting. One has only to think of the interest taken by the most diverse people in the microscopical discoveries of van Leeuwenhoek to realize how widespread is an interest in unfamiliar natural objects. When it was discovered by Abraham Trembley almost exactly two hundred years ago that an organism exists (we now call it *Hydra*) which feeds like an animal but buds like a plant, and reorganizes itself into two or more individuals if cut into bits with scissors, the interest aroused was such that polyps became, in the words of an anonymous eighteenth-century writer, "à la mode." [3] Interest in the unfamiliar is abundantly illustrated by the history of science. Even in modern times, when people tend to be less enthusiastic than they were two or three centuries ago, the discovery of a living fish belonging to a group thought to have been extinct for some sixty million years caused great excitement, and a popular weekly journal devoted a large double page entirely to the event.

Just as the unfamiliar attracts the interest of both layman and scientist, so also does the orderly. In a low form one sees the appreciation of the orderly exhibited in a collection of butterflies systematically arranged by a collector who understands little of the life-processes of what he collects. No sharp line of separation can be drawn between the simple arrangement of natural objects in an orderly fashion and the systematic presentations of natural knowledge by great scientists. I found this out many years ago when demonstrating to a class of students preparing for the final Honours examination in zoology at Oxford. We were studying the anatomy of certain marine worms, and I noticed that one of the women-students had a book beside her, open at a coloured plate showing the external characters of some of the animals that we were studying. The book was unfamiliar to me and I stooped down to look at it. The name gave me a surprise that I have not forgotten. I learnt a useful lesson in modesty that day, which I should be happy to share with any scientist who thinks himself a different kind of being from the layman. The student, preparing for the highest examination in zoology at a great university, was using *The Seashore shown to the Children*.

There is a widespread belief in the "worth-whileness" of finding out. The community as a whole appears to approve of the setting apart of a limited number of talented people for the express purpose of discovery, without requiring that all research should be directed towards material ends. The public expects as almost a matter of course that some one or other should concern himself with all branches of natural knowledge. This was forcibly brought home to me some years ago when I was one of the three or four people in the world who were making systematic studies of the causes of breeding seasons. When I remarked to non-scientific friends that the environmental causes which regulate the breeding seasons of animals were not known—that no one knew what makes the blackbird breed in early spring—I was met by frank incredulity. "Oh, *some one* knows," I was assured; "the experts *must* know." It seemed intolerable that a community which maintains people expressly for the purpose of getting all sorts of knowledge should not be able to obtain information on such a very straightforward and familiar subject.

There is one particular kind of knowledge which both the scientist and the layman place high up on the scale of values. This is the knowledge that throws light on man's place in the universe. The discoveries of Copernicus and Darwin caused a ferment of excitement which shook and changed the outlook of the whole civilized world, quite apart from any application to material human welfare. Again, one's whole outlook on the universe is changed and broadened by the knowledge that great groups of animals, some of them of gigantic bulk, have arisen in the distant past, evolved, persisted for millions of years, and then become totally extinct millions of years before man, or even his ape-like ancestors, appeared on earth.

5. *The Appreciation of Appreciation*

One of the values of science to the scientist is of a kind that is so generally understood in all fields of human activity that it is only mentioned here for the sake of completeness. The successful investigator appreciates the appreciation of others, provided that the others are qualified to judge. The extent to which scientists are affected by the desire for the approval of others varies widely. One of the greatest, Henry Cavendish, was so little affected by it that he did not bother to publish some of his most marvellous discoveries. Most scientists, however, naturally like their colleagues to think well of them, and they value science partly because it is an activity in which they can earn the approval of others.

6. *Has Truth Intrinsic Excellence?*

So far we have been considering value as equivalent to the existence of conscious appreciation. Another aspect of the value of science, attractive to the intuitions of many people, may be illustrated by an imaginary event.

Let us suppose that a group of psychologists, having armed themselves with a marvellous new invention which enables them to assess happiness objectively, accompanies a scientific expedition which sets out to explore two islands. When they analyse their data, they find that the average happiness or general contentment of the people on the two islands is exactly the same. Meanwhile the anthropologists of the expedition have been studying the natives' outlook on the universe. On the one island thunder is ascribed to the anger of the tribal ancestors, boiling springs are regarded as giants' cooking vessels, the birth of twins is regarded as indicating that the agricultural crops will be prolific, etc. On the other island these and other natural phenomena are interpreted in accordance with the scientific ideas with which we are familiar. In both islands the phenomena are regarded with interest, which is equal in the two cases. Which is the better civilization (apart from future prospects)?

There may be sceptics who will deny that a balance can be held between the two islands. Others may consider that if one island's civilization is better than the other, that is solely because some external observer appreciates the one civilization more highly, for in the absence of an external observer no difference exists. Most people, however, are likely to say that the civilization in which there is true knowledge is the better. Truth, in fact, has intrinsic excellence, apart from its effects. This belief—for it seems impossible to prove or disprove in any formal way the statement that truth has or is a value—has been a mainspring of scientific research, particularly plainly exhibited in the lives of such scientific geniuses as Charles

Darwin and T. H. Huxley, but animating also many much lesser men and women.

Diametrically opposite to these ideas stand those of the rulers of totalitarian states. Some general remarks on the subject are attributed to Hitler: "There is no such thing as truth. Science is a social phenomenon, and like every other social phenomenon is limited by the benefit or injury it confers on the community." [87] Himmler has applied these principles to a particular case, when attacking German scholars who refused to acknowledge the genuineness of a forged document on German archæology. It surprised him that anyone should make a fuss as to whether it were true. "The one and only thing that matters to us," he is reported to have said, "and the thing these people are paid for by the state, is to have ideas of history that strengthen our people in their necessary national pride." [87] As the Nazi professor of philosophy at Heidelberg announced, "We do not know of or recognize truth for truth's sake." [108] For Hitler and Himmler and the Nazi professor it seems nonsense to worry whether a given statement is true or not: the only thing that matters is how that statement affects the community. It is probable that few first-rate scientists would assent to what they regarded as an untruth, even if they could be persuaded that such assent would be materially beneficial to the community. It is apparent that the orderly structure and dependability of science would become transformed into chaos if Hitler's and Himmler's ideas were accepted by scientists as a whole; and scientists have always been accustomed to place a very high value upon truth, generally without considering the philosophical background of the position that they adopt.

7. *Propaganda fostering a Feeling of Disillusionment in Science*

Not only in Germany, but in the U.S.S.R. and Britain as well, movements have arisen which are in opposition to the opinion that science has a value as an end in itself. Although some of the writings of these schools contain so many contradictions that precise refutation is rendered impossible, yet there is no doubt that the intention is either to deny the value of science altogether apart from its purely material services, or else greatly to minimize it. Professor J. D. Bernal, for instance, sneered at science outside the Soviet Union as an "elegant pastime." [9] Elsewhere he writes of science (apart from its practical applications) as a "game." [8] There is no economic system, he says, which is willing to pay scientists "just to amuse themselves." Science, he says, "has all the qualities which make millions of people addicts of the crossword puzzle or the detective story, the only difference being that the problem has been set by nature or chance and not by man, that the answers cannot be got with certainty, and when they are found often raise far more questions than the original problem." [8] These, the reader should note, are said to be the *only* differences. Professor Bernal has attacked the high ideal of science as presented in T. H. Huxley's *Method and Results* and said it was a form of snobbery, "a sign of the scientist aping the don and the gentleman. An applied scientist must needs appear somewhat of a tradesman; he risked losing his amateur status. By insisting on science for its own sake the pure scientist repudiated the sordid material foundation on which his work was based." [8] All this is propaganda fostering a feeling of disillusionment in science.

A very interesting sidelight on the fostering of disillusionment in science is provided by Mr A. L. Rowse, Fellow of All Souls College, Oxford, in his recent book, *A Cornish Childhood*.[90] On pp. 174–175 he tells us that from the first he shared "the aversion which nearly all the most intelligent men I have met since have had for science." "I am not merely uninterested," he insists, "I am positively anti-scientist." He thinks people should learn science, if at all, "after they have been educated. . . . And of course I regard the tremendous cult made of science in modern society as very exaggerated: scientists have been very skilful in putting themselves across, and they have a vociferous and docile claque to support them in their claims." Modern society rests upon scientific foundations "in the sense that a modern town rests upon a proper sewage and drainage system." Having produced this propaganda of disillusionment in science, Mr Rowse goes straight on to say that he is "in entire sympathy" with a certain contemporary movement in science. The members of this movement, he says, are mostly friends of his. He mentions a string of six of them—and they are six of the seven leading figures in the movement for the central planning of science.

If the movement responsible for the propaganda of dis-illusionment in science could gain power, it would gravely damage science. Already its power is considerable, not only in the U.S.S.R. and Germany, but also in Britain. Influenced by the thoughtless desire to be considered modern and fashionable, a desire that is unaffected by the ups and downs of civilization, many young scientists are beginning to tread a course which, if it succeeds, will lead to the eclipse of their subject. The situation is the more threatening because nowadays, as a result of the expansion of science, there is a far higher proportion of scientists who are not enthusiasts than in previous centuries. It will be suggested in Chapter V that the first social responsibility of scientists is to oppose the threat.

Chapter III

FREEDOM OF INQUIRY

"The real scientist . . . is ready to bear privation and, if need be, starvation rather than let anyone dictate to him which direction his work must take."—SZENT-GYÖRGYI.

1. *The Nature of Free Inquiry*

FREEDOM of inquiry means the freedom of the research worker to decide what he will investigate. It has been said contemptuously that such freedom would enable a man to spend his life, at the expense of the community, in proving that the earth is flat. This idea must be contradicted. The research worker who does not do good research betrays his trust, either through lack of gift or as a result of laziness. The value of his research can be assessed only by his fellow-scientists. Their judgment may be wrong, but no better criterion of excellence exists. No scientist should be paid to continue indefinitely with trivial or negligible research. He must not, of course, be hurried to produce results—that would be fatal; but if, over a period of four or five years, he is consistently unfruitful, facilities should be given him to engage in some scientific activity other than research. Some one else will then be free to take his place.

Let it be clear that freedom of inquiry does not and cannot mean perfect freedom. A scientist who wants to work at the tree-top fauna of the Brazilian forest can generally be free to do so only if he can persuade other people that his study is likely to be fruitful: then and then only will funds be forthcoming. Similarly, scarcely anyone is free to have a cyclotron or an ultra-centrifuge at his disposal unless he can make other people believe that his use of such an expensive instrument would result in important advances in science. An enormous amount of scientific work can be done, however, with little expenditure, and in so far as his imagination directs him towards relatively inexpensive studies, the established research worker in a British university is generally remarkably free.

Aldous Huxley has made one of his characters say that the real charm of scientific research is its easiness. The exact contrary is true. A life of free research is very strenuous, because the investigator must continually be making decisions on matters in which he has only his intuition to guide him. It is generally recognized that it is hard and anxious work to make decisions with inadequate information, and all the more so if the result of the decisions determines the success or failure of the person who makes them. The investigator must continually be deciding what he will do next in the laboratory and what he will read in the library, and how much time he will spend in each. Continually new possibilities of investigation suggest themselves to him as a result of what he does or reads. A wrong decision may waste months, a right one may bring great success. The life is too strenuous for most people, and the timid scientist hankers after the safety of directed team-work routine. The genuine research worker is an altogether different kind of person. He gets the ideas on which scientific research depends at the most extraordinary and unpredictable times, seldom when he is expecting them. If the idea is a good one, a complete change of research plans may be necessary. It may take months to discover whether it is a good idea. He quietly weighs the risks. If he considers the chance good enough, he throws over what he is doing and enters with enthusiasm into the new investigation. He will gain or lose according to the correctness or falsity of his judgment. Only a man of courage, independence, and lively imagination will take big risks for the possibility of big discoveries. "I am like a gambler," wrote Charles Darwin, "and love a wild experiment." [23]

The great biochemist, Szent-Györgyi, has written these words: "What I want to stress is that the pre-condition of scientific discovery is a society which does not demand 'usefulness' from the scientist, but grants him the liberty which he needs for concentration and for the conscientious detailed work without which creation is impossible. . . . The real scientist . . . is ready to bear privation and, if need be, starvation rather than let anyone dictate to him which direction his work must take." [98] These are the opinions of the man who was the first to isolate a chemically pure vitamin. He is a Nobel Prizewinner for Medicine.

Faraday administered a most effective and at the same time courteous rebuke to a would-be planner who wanted to make him continue some technological researches on glass. He said that as he had been obliged "to devote the whole of my spare time to the experiments [on glass] already described, and consequently to resign the pursuit of such philosophical inquiries as suggested themselves to my own mind, I would wish, under present circumstances, to lay the glass aside for a while that I may enjoy the pleasure of working out my own thoughts on other subjects." [30] Instead of improving glass, Faraday went on to the researches in science that made him one of the greatest investigators the world has ever known and revolutionized industry as well.

2. *"Overlapping" in Research*

When there is freedom of inquiry, two or more research workers often do the same sort of work independently at the same time. People argue that if the whole of scientific research could be comprehended under a central organizing authority, this overlapping could be avoided. The belief that it is obviously desirable to avoid overlapping must, however, be contested. It is necessary to enter in some little detail into this seemingly simple matter.

Every research worker needs a large store of *judgment*. A constant stream of scientific papers is flowing forth from all parts of the civilized world, and the scientist's decision as to what he shall investigate depends on what he believes in those papers. It would be ingenuous to imagine that each paper contains nothing but the truth.   Papers contain errors of fact, slips and misinterpretations, and the scientists who write them often deceive themselves by adherence to fashionable hypotheses.   The scientist takes nothing as true on anyone's authority, for he knows that every investigator may make mistakes.   Nevertheless, he cannot check every statement in every paper he reads.   If he tried to do so, he would spend his life on nothing else—and would be able to read very few papers.   He has got to be able to form a judgment of what is true.

His judgment is helped when one scientist repeats the work of another and publishes his results.   There is one occurrence, however, which helps the scientist to form a valid judgment better than anything else.   This is the simultaneous or nearly simultaneous publication of the same result by two entirely independent workers. Central planners are inclined to consider that one of the two independent workers has been wasting his time.   The actual research worker knows that this is not so.   It is the very fact that the two workers are independent that inclines others to accept their findings. Scarcely a working scientist will deny that two independent papers containing the same result are very much more convincing than a single paper by two collaborators.   Independence has another advantage.   If the two had been working together, their minds would have tended to run along the same track.   Since they did not, each paper has a different outlook, and the reading of the two papers is far more stimulating and suggestive.   An excellent example is provided by the almost simultaneous publication of the two papers that founded our modern knowledge of valency (see p. 49).

Some degree of isolation from other scientists is conducive to originality.   There is an optimum amount of contact, which should not be exceeded.   Anyone who doubts this should reflect how damaging to research it would be, if the teaching of science at all universities throughout the world were exactly the same.   Each scientist can hold only a small fraction of the available information in his head, and it is desirable that different scientists snould think different thoughts. It is obvious that one can spend too much of one's time in the library, and single-track-mindedness is even more contagious by personal contact than infectious through print.

3. *Chance in Research*

The free investigator makes his own plans for research, but does not keep to them inflexibly. He has the courage and determination to throw away the toil of months if he gets an idea which he judges likely to lead to greater results. He is always alert and ready to switch his energies in a new direction when the unexpected appears. Chance may offer him a great new opportunity. The element of chance has always been important in scientific research, and only by free inquiry can the utmost use be made of it.

One of those who believe in the central planning of science [21] has said that freedom was useful to the science of the eighteenth century, but is not suited to the twentieth. I therefore select an investigation that is still in active progress to-day, as an example of the way in which free science still produces great results in modern times. The investigation is of high intrinsic and practical interest. The story of the discovery of the new therapeutic agent, penicillin, illustrates in a remarkable way how unpredictable the course of discovery is, and how free science produces results which could not be centrally planned.

In 1929 Dr A. Fleming,[32] working at St Mary's Hospital, London, left a number of culture-plates of staphylococci exposed from time to time to the air. The plates became accidentally contaminated by various micro-organisms. On one plate there grew a mould, *Penicillium notatum*. Dr Fleming noticed that round that mould the staphylococcus colonies died. The mould produces an anti-bacterial substance. This was not in itself a novelty, for it had been known for a long time that one micro-organism sometimes inhibits the growth and multiplication of ,another.   Dr Fleming filtered a broth culture of *Penicillium*, and called the filtrate penicillin. He made practical use of this substance in separating *Hæmophilus influenzæ*, which is resistant to it, from other bacteria.   He suggested the use of penicillin in the local treatment of infected wounds.   Unsuccessful attempts were made to isolate the bacteriostatic substance, and the subject lapsed from immediate interest.

In 1939 it occurred to Professor H. W. Florey and Dr E. Chain, of the School of Pathology, Oxford, that it would be profitable to conduct an investigation of the chemical and biological properties of the anti-bacterial substances produced by bacteria and moulds.   They decided by great good fortune to start work with Fleming's mould.   The results are well known.[16]   Penicillin is a bacteriostatic agent with properties that can only be described as astonishing.   It prevents the multiplication of certain bacteria at a dilution of one part in fifty million, yet it is harmless when injected into the body at high concentration.   As Dr Chain has himself said to me, it would have been absolutely impossible to plan a research to find such a substance, because the existence of such a substance was not envisaged by anybody.

It is necessary to stress the great element of chance in this investigation.   *Penicillium* came by chance to one of Dr Fleming's cultures, when he was not studying the influence of one micro-organism on another.   That is only a small fraction of the element of chance in this case.   Hundreds of species of *Penicillium* have now been examined, and penicillin has been found only in *Penicillium notatum*, the species which alighted on Dr Fleming's culture.   That again is by no means all.   There are sixty strains of *Penicillium notatum*. and only one of these—Dr Fleming's—produces penicillin. Again, penicillin may be the only really valuable antagonistic substance produced by moulds or bacteria, and Dr Chain and Professor Florey might have chosen one of about twenty others for their study.

In stressing the action of chance, which I believe to play— now as in the past—a very large part in discovery, I wish to make it clear that chance does not detract from the credit due to Dr Fleming, Professor Florey, and Dr Chain for their remarkable discoveries. As Pasteur himself said, "Chance only favours prepared minds." [106] Lagrange said almost exactly the same thing: "Accidents only happen to those who deserve them." [104] Reaumur had already said it even more precisely when Lagrange was yet only six years old. He was referring to the wonderful discovery of regeneration made by Trembley when the latter had chanced to find *Hydra* in some water collected in a ditch. "Chance alone," wrote Réaumur, "has given the opportunity for a discovery to be made which reason scarcely permits one to believe [even] after one has seen it; but it was one of those chances that only offer themselves to those who are worthy to have them, or rather, to those who know how to procure them." [88] These are thoughtful words, and as applicable to-day as they were two hundred and two years ago.

When once Professor Florey and Dr Chain realized that they had made a great discovery, it was a straightforward matter to plan how it should be followed up. It was necessary to find a way of culturing the mould on a large scale; to devise standards for assaying it; to get it as pure as possible and attempt analysis of the active principle; to carry out fuller tests of harmlessness and bacteriostatic activity; and to try its administration to man. [1] This is exactly the sort of situation in which organized team-work is valuable: the primary discovery has been made, and the follow-up can be planned. A team of workers co-operating with the original discoverers promises to give the world one of its most important therapeutic agents (Abraham, Chain *et al*, 1941). At present the chief requirement is a knowledge of how to increase the yield of penicillin, or how to synthesize its active principle. Meanwhile, the results of its use are already very encouraging, indeed astonishing, in the treatment of a wide variety of illnesses.

4. *The Contribution of Solitary Workers to Twentieth-century Science*

Professor J. D. Bernal has said that "practically the whole of the great advances of science in the twentieth century were achieved not by scientists working as individuals, but in organized groups." [9]   Mr Swann has said almost exactly the same thing: "Only a small proportion of the scientific discoveries made in this or any other country are due to individuals working on their own." [97]   Statements of this kind tend to be repeated and gain credence. It is easy to refute them. In the first place, when two or three scientists work and publish together (in this century as in the last), they cannot, in the majority of cases, be said to form an organized group. No one has coerced them into working together. They find it convenient, for a special purpose, to collaborate. That is a very different thing from organized team-work, such as that in force at the Physico-Technical Institute at Kharkov, in the U.S.S.R., where the investigators are organized in brigades, and no one is allowed to start working on a new problem without permission. [18]   That is an organized group, and one can say confidently that such organized groups have made only a relatively small contribution to discovery.

The second part of the answer to Professor Bernal's statement shows him even more clearly to be wrong. Here, there is no question of the interpretation of terms. He says unequivocally that single investigators have contributed almost nothing to the great advances of science in the twentieth century. Let us consider that.

What about Einstein? Is it really possible to forget about him? "I am a horse for single harness," wrote Einstein, "not cut out for tandem or team-work." [54]   This was the man who suggested jobs as lighthouse-keepers for refugee scientists, so that they might have the isolation necessary for scientific work.

Einstein alone demolishes Professor Bernal's statement, but

no one must be allowed to think that Professor Bernal forgot about Einstein and was in other respects well informed. In many branches of science the fundamental discoveries have been made by scientists working as individuals, in this century as in others. The following are a few examples chosen at random from various sciences.

Probably the most fundamental problem in chemistry is the nature of the forces that hold atoms together, and in particular the nature of the bonds of valency. Our understanding of this problem dates from 1916, when two papers were published independently, the one in March in Germany, the other in April in the U.S.A. Although some preliminary ideas had already been published by Sir J. J. Thomson and Sir William Ramsay, yet Walther Kossel's "Über Molekülbildung als Frage des Atombaus" [62] and Gilbert N. Lewis's "The Atom and the Molecule" [66a] are the foundation of modern knowledge of valency. Both were workers who published under their own names alone. This does not mean that they lived like Trappist monks, and Lewis mentions the discussion of certain aspects of his subject with his colleagues; but our understanding of the nature of valency is due to the genius of these two independent workers. Kossel concentrated on electrovalent and Lewis chiefly on covalent linkages (without at the time appreciating the difference). Three years later, when Irving Langmuir published his well-known paper on "The Arrangement of Electrons in Atoms and Molecules," [66] he began, necessarily, by full references to the work of Kossel and of Lewis. Every chemist will allow that the fundamental work leading to the modern theory of valency was done by scientists working and publishing separately.

The same applies to the fundamental work in cytogenetics. The close parallel between the results of Mendelian breeding experiments and the behaviour of the chromosomes was first pointed out by a scientist, W. S. Sutton,[96] working as an individual. Sutton's paper constituted the actual origin of

D

cytogenetics. The next most important step was the production of evidence that the sex chromosomes are concerned with sex determination. This was done by C. E. McClung,[71] who was also working as an individual though he directed his students to similar studies. McClung's fertile work was followed by that of the great cytologist, E. B. Wilson, who first established the true facts about the relationship between the sex chromosomes and sex determination in a series of papers written by himself alone. [110]

We may turn to any part of science and we are likely to find the same thing: the fundamental discoveries are commonly made by single workers. The elusive Golgi bodies, so long undetected but now known to be a component of almost every animal cell, are nearly always made visible by one of two methods: either by their capacity to reduce osmium tetroxide very slowly, or else by the so-called photographic method, in which silver nitrate and a photographic developer are used. The osmium tetroxide method was discovered by a solitary worker (F. Kopsch), [61] and so was the photographic (by S. Ramon y Cajal). [86]

In the wide and intensely interesting subject of experimental embryology, to take another example, almost all the fundamental twentieth-century discoveries have been made by single workers. This is the subject which concerns itself with the experimental analysis of the causes that make an apparently simple egg develop into an excessively complicated adult organism. The organ-forming potentialities of the various parts of the egg were described in the well-known papers of the American, E. C. Conklin.* W. Vogt, working alone, introduced the method of staining small parts of the living egg and thus following those parts during development. His maps of the presumptive regions of the eggs of Amphibia, made in this way, are famous. The descriptive method introduced by Vogt has thrown a flood of light on experi-

---

* References to the work of the experimental embryologists mentioned may be found, *e.g.*, in Huxley and de Beer.[62]

mental studies in embryology. The concept of axial gradients or gradient-fields was developed independently by T. Boveri in Germany and C. M. Child in America—both single workers. The momentous discovery of the "organizer" is due to the solitary work of H. Spemann, who, although he collaborated with Mangold in a well-known paper, yet had previously discovered the essential principle in independent research. Finally, that classical work *On Growth and Form* is the achievement of the single mind of D'Arcy Thompson, [101] though he acknowledged the informal help of his friends. I believe that scarcely any biologist will deny that the separate and solitary work of Conklin, Vogt, Boveri, Child, Spemann, and D'Arcy Thompson has been, in this twentieth century, of altogether outstanding importance.

The study of vitamins tells the same story. The concept of a deficiency disease is due to the Dutchman, G. Grijns,[38] who suggested at the beginning of this century that a diet containing sufficient proteins, carbohydrates, fats, salts, and water might fail to sustain life.* This was directly contrary to the scientific opinion of the day. Grijns's collaborator, E. Eijkman, was influenced by the prevailing opinion, and it was not until Eijkman had returned from the East Indies to Holland that Grijns was able to arrive at his astonishing conclusion. Grijns was working at what we now call vitamin B1 or thiamin, and this substance was first obtained in (impure) crystalline form by another solitary worker, C. Funk.

Hormones tell the same story again. Thyroxine, for instance, was first isolated by E. C. Kendall,† and the parathyroid hormone by J. B. Collip. J. Takamine gave the first satisfactory method of extracting adrenaline. A. Butenandt was a pioneer in the isolation of sex hormones.

The story is endless. Almost every scientist can continue it. Fleming's discovery of penicillin is an example. Professor

---

* Rigorous proof of this fact was given later by F. G. Hopkins.
† For references to work on the isolation of hormones, see **Harrow and Sherwin**.[44]

Bernal's statement that "practically the whole of the great advances of science in the twentieth century was achieved not by scientists working as individuals, but in organized groups" is contrary to the demonstrable facts. The solitary worker has made an enormous contribution to twentieth-century science.

The solitary worker owes much to others—to all those who gave him encouragement as a child or at any time, to his teachers in school and university, to the authors of papers and books on his subject, to colleagues who chat with him, to architects and engineers who provide him with a laboratory, to editors, publishers, and printers who make his discoveries known, to laboratory assistants who give him practical help. The community helps him: he makes return primarily by the discovery of demonstrable facts. He finds that he makes the best return when he directs his own research and carries it out as he thinks fit. He desires to dominate no one and to dictate to no one how research shall be carried out, but he considers that in the best interests of science he, and others who feel as he does, should be free to work alone.

5. *The Place of Teams in Research*

In strong contrast to directed team-work stands the co-operation of two or three scientific friends. Such co-operation has produced a great deal of excellent scientific investigation. It was the friendly collaboration between Schleiden, the ex-lawyer and botanist, and Theodor Schwann, the anatomist, that brought the cell theory into existence.[33] The two men, both in their thirties, were dining in Berlin in October 1838. Schleiden described to Schwann the nucleus of plant cells. Schwann at once recollected that he had seen a similar structure in cells of the Vertebrate spinal cord. The two men went forthwith to the University Anatomical Institute, where Schwann carried out his researches. Schwann showed Schleiden the cells of the spinal cord, and Schleiden at once recognized the nuclei as the same structures as those with which he was familiar in plants. Although a good deal was already known about cells from the pioneer work of Dutrochet, Turpin, Brown, von Mohl, Purkinje, and Müller, [17, 93] yet this occasion may justly be regarded as that to which we owe the first general formulation of the cell theory. The two men were not coerced to work together: in fact, they worked and published separately. Although they made big mistakes, yet their work served to focus attention on the cell and to show how it forms the basis of the structure of both plants and animals.

The sort of informal intercourse that Schleiden held with Schwann, as well as closer collaboration resulting in the publication of joint papers, has long been of great importance in discovery. One of the most valuable kinds of scientific collaboration is indeed entirely informal. Every investigator could illustrate this from his private experience. For instance, I have never had a formal conference with either of the two men who occupy the research-rooms next to mine, but we naturally meet accidentally in the passage and exchange a few remarks, or drop in on one another to ask a question; and it would be hard to exaggerate the benefit that my research has gained from friendly help given in these chance encounters.

People who like to control everything under an orderly plan cannot believe that independent research and informal collaboration are more *efficient* than directed teams. It is urgent, for the sake of the welfare of science, that people who worship mere tidiness should occupy themselves in some suitable and congenial occupation and not strive to impose impossible conditions of work on the original scientific investigator, whose mind they can never hope to understand. The would-be central planner of science is a menace to progress. The Archbishop of Canterbury has said, "There is a real danger of planning in this country being carried out for the wealth or the power of the planners." [15] We cannot tell whether the planners want wealth or power or only tidiness, and their motives are private and irrelevant; but the Archbishop has done well to point out the danger, which threatens science as much as any other human activity.

The proper function of a research team is to work out the consequences after an independent worker or two or three scientific friends have opened a new line of investigation. There will be plenty of people who want to follow the new line. Indeed, one notices a strong tendency for scientists to ask, "What is being done?" They might as well ask frankly, "What is the fashion?" The original investigator on the contrary asks himself, "What is *not* being done?" The people who want to *follow* a new line often do excellently in teams and they can be fitted satisfactorily into planned research. They have neither the wish nor the ability to think originally, though they are often talented, well equipped technically, and possessed of a great love of knowledge. If science is to flourish, however, encouragement must be given to people of independent spirit who want no master. The desire to know is widespread among men: the desire to know specifically that which is not known is on the contrary very rare.

Although teams have their place in scientific research, yet they also have grave disadvantages. As the director of a research team once said to me, "The trouble with a team is that directly a man has joined it, he doesn't bother to have ideas any more." The statement was an exaggeration, but it contains an important element of truth. If the investigator works in a team, it is difficult for him to allow ideas to grow in his mind. He knows that his colleagues will frown upon his wish to give up what he is doing, for that will disturb their work. He knows also that instead of throwing himself into the congenial task of entering immediately on the new work, he will have to devote most of his energy to the very uncongenial task of trying to persuade the director of the team to allow him to upset all the carefully prepared plans. His position is exactly what that of a composer would be if he had to get other people's agreement before he could work out in detail the themes that occurred to him in moments of inspiration.

The free research worker is never on holiday. He must always, day and night, be ready for the arrival of an idea. He can only achieve that readiness by so arranging his life that ideas may come. He cannot order them to come, but he can and must provide an environment in which their coming is a possibility. For this some solitude and mental quiet are necessary. He lives for his research not only when he is in the laboratory or library, but right round the clock. My friend Dr V. Korenchevsky was present when Pavlov, the famous Russian physiologist, was asked for a recipe for success in scientific research. His advice was simple. Get up in the morning with your problem before you. Breakfast with it. Go to the laboratory with it. Eat your lunch with it. Take it home with you in the evening. Eat your dinner with it. Keep it before you after dinner. Go to bed with it in your mind. Dream about it.

It is impossible to live for ideas when one knows that their coming will simply mean a public contest as to whether they shall be tried out. Until an original idea has been tried, its very originality ensures that it is something *unlikely* in the existing state of knowledge. Ideas are rightly the personal property of their conceivers until they have been tried. When they have been tested and proved, the scientist has a duty to publish them, for there should be no private property in demonstrable knowledge; but the original idea is something personal and private until its truth or falsity has been demonstrated.

The picture of the great scientist sitting in his study throwing off ideas for his vassals to work out in the laboratory is fantastic, for two reasons. First, no matter how great a scientist may be, worth-while ideas are definitely rare, and are most likely to come if the scientist spends a good part of his time in manual work in the laboratory. Secondly, every vassal who carries out the great man's ideas is by that very fact precluded from carrying out his own, and so he will seldom "bother" to have any.

The team is an excellent institution for carrying out an obvious bit of work, provided that it does not take an original mind away from free research; but it is not an organization for producing ideas. It is a group of specialists, each devoting his special knowledge to one side of a problem; and there is always grave danger that no one is seeing the problem as a whole. The director himself is likely to be too much occupied with administrative work, too eager to seize every opportunity to rake in some more funds for the research, too energetic in making all the right contacts, too liable to prefer the committee-room to the laboratory to have the mental repose necessary for true originality. If a man is good enough in his subject to be the leader of a team, he is likely to be wasting his talent; for he may be good enough to have the privilege of being allowed to work alone.

6. *Cancer and Freedom of Inquiry*

The planning enthusiast would deny freedom to the research worker because he thinks central direction more efficient. In my last book, *The Scientific Life*, [6] I pointed out (on p. 75) that Röntgen and the Curies, who helped sufferers from cancer more than anyone else ever has before or since, did not engage in research on cancer. On the contrary, they investigated X-rays and radium, without thought of malignant tumours. They were simply very talented people who were allowed to carry out their own experiments for the purpose of increasing knowledge. "It may be anticipated," I continued, "that when another great discovery comes to bring help to sufferers from this disease it is likely to come from an equally unexpected quarter." It is an amiable foible of the human race to like being able to say, "I told you so." Not being an exception to the rule—who, indeed, except Mr Winston Churchill, can claim to be that?—I cannot forgo the pleasure.   What I said has come true.

The newspapers have splashed the information about the control of prostatic cancer.   The facts are certainly impressive. American statistics show that about five per cent. of men over fifty years of age die of cancer of the prostate gland.   This terrible disease is extremely painful, emaciates the patient and may even paralyse him; it leaves him without appetite or the desire to live. [25]   A method of controlling this disease has now been found, which is often successful.   It involves no operation.   "To see these symptoms removed," says the Editor of the *Lancet*, "appetite for food and life restored, and the patient once more a useful citizen is a dramatic experience." [25]   Dr Charles Huggins, who has had so much to do with the discovery of the new method of treatment, has written as follows: "It is striking to see patients emaciated from malignant disease develop a voracious appetite . . . pain often disappears. . . . The increased food intake and decrease of pain promote a sense of well-being and more tangibly a gain in weight and increased blood formation so that the anæmia accompanying the tumour frequently disappears." [49] What is the origin of this marvellous discovery?   The newspapers, of course, never tell us that: they give all the credit to those who fasten the last link in the chain.   The discovery has its roots in the eighteenth century and owes its existence to two separate departments of knowledge, both apparently unrelated to cancer and also to each other.

The prostate gland is a male accessory organ of reproduction, situated near the base of the urinary bladder.   Its function is to add a secretion to the semen,* favourable to the activity of the spermatozoa.   To trace the history of the control of

* For the sake of any boy or girl reader who may be unfamiliar with the word *semen*, I mention that this is the fluid passed from male to female at sexual union. It contains many microscopical male germ cells or spermatozoa, one of which fuses with the female germ cell or egg, which is also microscopical in the case of mammals (including human beings).   The product of fusion grows and becomes the embryo.

cancer of this organ it is necessary to go right back to that grand old eighteenth-century physiologist and surgeon, John Hunter; for it was he who first noted that the prostate of a castrated animal (a bullock, to be precise) is small and flabby and contains little secretion. It was not until 1847 that Dr W. Gruber,[40] of St Petersburg, made the same discovery for man. Subsequent workers noted the same fact from time to time, and at last, in 1889, a serious study of the relationship between the testes and the prostate was published by Joseph Griffiths,[37] of Cambridge University, who had investigated the seasonal changes in size of the prostate gland in the mole and hedgehog and noticed the effect that castration has on that organ in the dog and cat. As we should say now, the testes produce a hormone or chemical messenger which circulates in the blood and stimulates the prostate to grow and secrete. When the testes are removed, the prostate declines. In 1926 Steinach and Kun[94] showed that injection of the female hormone into male animals also causes a rapid decline of the prostate gland.

That is one of the roots of our modern knowledge. The other is in biochemical studies having, at first, no connexion whatever with the prostate gland or with cancer. It was in 1912 that Grosser and Husler,[39] working at a children's clinic in Frankfurt, first discovered *phosphatase*. They discovered it in the lining membrane of the intestine. Phosphatase is a ferment capable of releasing phosphoric acid from certain of its compounds. The next discovery of importance was made almost simultaneously but independently in 1934 by D. R. Davis[24] at the Welsh National School of Medicine at Cardiff and by Baaman and Riedell[5] at the Technical High School at Stuttgart. These workers discovered that there are in fact two different phosphatases, the one active in acid and the other in alkaline media. Next year Kutcher and Wolbergs,[63] of the Physiological Institute of the University of Heidelberg, made the discovery that was to link together two lines of research. They found that human prostatic secretion is very rich in "acid" phosphatase, that is, the phosphatase that is active in acid media. (The significance of the presence of a relatively high concentration of acid phosphatase in the secretion of the prostate is irrelevant to our discussion.)

Our modern knowledge of how to control cancer of the prostate is due to the researches of these men—of Hunter, Gruber, Griffiths, Steinach, and Kun on the prostate; of Grosser, Husler, Davis, Baaman, and Riedell on phosphatase; and of Kutcher and Wolbergs on phosphatase in the prostate. *Not one of these men was studying cancer*, yet without them the discovery of the new treatment would not have been made.

Now at last we come to cancer itself. A. B. and E. B. Gutman, [41] of New York, discovered in 1938 that the *blood* of patients with cancer of the prostate often contains more than the normal amount of acid phosphatase. Malignant prostate cells retain the capacity to produce phosphatase shown by the normal cells of that organ, and pour it into the circulation. High amounts of acid phosphatase constitute a symptom of the disease and a means of diagnosis.

It was this knowledge that put Dr Charles Huggins, [49] of Chicago University, on the path that led to success   It was he and his associates who connected the knowledge of the hormone control of the prostate with the knowledge of the production of phosphatase by that organ.   It occurred to them to try removing the testes of some patients and supplying the female hormone to others, to check the secretory activity of the malignant cells and thus to reduce the amount of phosphatase in the blood.   Not only did they succeed in controlling the symptom: they controlled, in many cases, the disease itself.   The rest of the story has been reported in the newspapers.   Cancer of the prostate can in many cases be controlled by the supply of female sex hormone or of synthetic substances with similar physiological effects.

What central planner, interested in the cure of cancer, would have supported Griffiths in his studies of the seasonal cycle of the hedgehog, or Grosser and Husler in their biochemical work on the lining membrane of the intestine?

How could anyone have connected phosphatase with cancer, when the existence of phosphatase was unknown? And while it was yet unknown, how could the man in charge of the cancer funds know to whom to give the money for research? How lucky it is for sufferers from cancer of the prostate that Griffiths and Grosser and Husler and the others were not doing cancer research!

The linking together of a toe-nail and an umbrella by a surrealist is scarcely more unpredictable than the linkings that result in big discoveries in science, and no planner could make the right guesses. The basic knowledge is what matters most of all. It is mostly made by people whose names do not appear in the newspapers and who get little financial support for their work. Who cares about the people who first gave us knowledge of *Penicillium*, for instance, and taught us to separate its species and varieties? Or who cares, indeed, about the people who first started studying moulds at all? They were probably regarded as cranks by the ordinary people of their time. The millionaires spill their money on clinical medicine rather than on the basic sciences that make clinical medicine a possibility. Luckily the research workers in the basic sciences will continue their research whether they have ample funds or not, because it is their life-interest. The real research worker does not refuse to find things out simply because no one will give him a cyclotron or ultra-centrifuge costing many thousands of pounds. It is not for him to demand more money, though he could use it. One thing, though, he needs. Let the worker in clinical medicine have a grand laboratory, by all means, and a large salary and whatever else he wants. But to the investigator in the basic sciences grant *freedom of inquiry*.

Chapter IV

SCIENCE UNDER TOTALITARIANISM

"It is possible to defend the false bases of Mendelism only by lies . . . the teaching of Mendel and Morgan I cannot call anything but false."—ACADEMICIAN T. D. LYSENKO.[60]

"When he grasps Bolshevism, the reader will not be able to give his sympathy to metaphysics, and Mendelism definitely is pure, undisguised metaphysics."—ACADEMICIAN T. D. LYSENKO.[60]

"In order to get a particular result, one must want to get exactly that result; if you want to get a particular result, you will get it."—ACADEMICIAN T. D. LYSENKO.[60]

"If young people are forced to study genetics, their ability to think will be destroyed."—M. B. CHERNOYAROV.[60]

1. *The Growth of the Movement against Freedom in Science*

IN Britain, only about a dozen years separate us from the birth of the threat to freedom in science. An International Congress on the History of Science was held in London in 1931. A Soviet delegation, headed by Bukharin, attended. It was Hessen who expressed most explicitly the views represented by the delegation. He sought to show that Newton's researches were merely a by-product of the social and economic conditions of his country. It is not necessary to answer this contention in detail. Professor M. Polanyi [83] has reminded us that the same discoveries are sometimes made nearly simultaneously in countries where the social and economic conditions are radically different. During the nineteen-twenties those conditions were about as diverse as could be in India, the U.S.S.R., and Central Europe. The discovery of the Raman effect gained the Nobel Prize for the Indian, C. V. Raman, but it was independently discovered in the U.S.S.R. by Landsberg and also anticipated on theoretical grounds by the Austrian, Smekal, so that the term "Smekal-Raman effect" is sometimes used.

Despite the force of the criticism that can be directed against the views expressed by the Soviet delegation, Bukharin and his associates sowed a seed which germinated. Two years later, at the Leicester meeting of the British Association, the germination of the seed began to become apparent: the attention of scientists began to become diverted away from science towards social and economic questions. The change was apparent in the Presidential Address by Sir Frederick Gowland Hopkins. [48] As he himself said, "You may feel that throughout this address I have dwelt exclusively on the material benefits of science to the neglect of its cultural value." In his final sentence, however, the great biochemist boldly stated his belief in the value of science as "one of the Humanities; no less."

The movement against the pursuit of science for its own sake and against freedom in the practice of science suddenly began to become influential in this country in 1936, when the economist, Sir Josiah Stamp, gave the Presidential Address to the British Association at Blackpool. Sir Josiah recommended a "wise central direction" to allocate research workers to their tasks. [92a] *Nature* followed the lead and the movement grew steadily. Science began to be regarded as simply the product of the material conditions of ordinary human life, and as existing only to affect those conditions. People began to claim that scientists cannot be left free to choose their own subjects for research, but must submit to central planning so that their work may be devoted to material human wants.

To fortify the British Association in its new movement, a special Division was formed in August 1938, the Division for the Social and International Relations of Science. This Division held a three-day meeting at the Royal Institution, London, in September 1941, throughout the whole of which I was present. The new movement was much in evidence at this meeting and there was a lot of political propaganda. The culminating event was the reading out by the President, Sir Richard Gregory, of *The New Charter of Scientific Fellowship*, consisting of a preface and seven *Scientific Principles*. "The basic principles of science," he read out as the fourth principle, ". . . are influenced by the progressive needs of humanity." In this sentence, solemnly stated at the culminating moment of the conference and subsequently printed in *Nature*, [36] we have an approximation to the totalitarian idea of science. It is the very essence of true science that its basic principles are not affected by the needs of humanity. Those basic principles are the free search for demonstrable truth and the formulation of generalizations covering the discoveries made. The needs of humanity do not change them. It is only under totalitarianism that such a thing can happen. Let us recall Hitler's outlook on science (p. 38). "Science is a social phenomenon," he is reported to have said, "and like every other social phenomenon is limited by the benefit or injury it confers on the community. The idea of free and unfettered science . . . is absurd." [87] Let us recall also how Himmler thought it absurd for anyone to worry whether a document on German archæology were true or false, for the only thing that seemed to him to matter was whether it stimulated national pride. These sentiments are reflected in the British Association's fourth principle.

After the conference, people objected to the fourth principle and it was rewritten. The words "basic principles" were omitted and reference was made instead to the "structure" of science. [13] The obnoxious idea was thus replaced by a truism, for everyone knows that there is an interaction between science and industry. It has been claimed that the wording of the original statement was simply the result of careless drafting. It is likely that carelessness had much to do with the acceptance of the draft by the Council of the Association, but it is difficult to believe that some one on the drafting committee proposed the words "basic principles" when he did not mean them.

2. *War and Opinion*

It is true that totalitarianism was already beginning to have a bad influence on German science, when the outbreak of war upset research in every country. Nevertheless, so high was the level of scientific culture in Germany that a sort of momentum carried it forward and prevented the political regime from interfering very seriously, apart from the expulsion of brilliant Jewish investigators. If peace had been preserved, there would probably have been a progressive deterioration in German science while the Nazis remained in power. In Russia there was never much momentum of scientific culture. Russian science as a whole was never comparable with German, despite the influence of a few distinguished men. When totalitarianism was applied to Soviet science, an opportunity was therefore given to observe its effects, undisturbed by any very powerful influence persisting from earlier times. To anyone interested in scientific method the experiment provides a valuable object-lesson, from which conclusions can be drawn as to the conditions that favour scientific research.

A study of Soviet science does not support the view that totalitarianism is favourable to science. It might seem churlish to criticize the institutions of our ally, when we all know that her action in self-defence has made our task in the war so much lighter. It would indeed have been congenial, even in a book devoted to the subject of the relation between science and totalitarianism, to omit all criticism of our ally, on account of the benefits that have accrued to us from the development and use of her military might.

Nevertheless, scientists who are supporters of the Soviet political regime have created a situation that makes it desirable, indeed necessary, that criticism should be offered. By pouring out a stream of praise for Soviet science in books, articles, and addresses they have created a body of opinion that could not have been formed if those who disagreed with them had felt free to answer. Now at last an answer must be made, for the swing of opinion, fostered by one-sided argument, threatens to have serious results for the science of democratic countries. If anyone censures the publication of adverse criticism of the science of our ally, let him direct that censure against those who have traded on the unwillingness of others to offer such criticism. Truth cannot come from one-sided argument. Nothing but freedom of speech and publication can reveal it.

It cannot be right to praise the science of another country simply because that country is our ally. Good science is to be respected wherever it comes from; bad science, neglected or condemned. We rightly hear and commend the great music of German and Austrian composers. That we fight those countries is irrelevant. Similarly we should commend all that is good in the science of those countries. In commendation and censure our judgment should be unaffected by circumstances of alliance or hostility.

When the U.S.S.R. attacked Finland there was no outburst of praise for Soviet science in Britain. When Stalin made a pact with Hitler, *Nature* did not print columns of praise for the science of the U.S.S.R. But when Hitler attacked and thus made Britain and the U.S.S.R. allies, all those who believe in the central planning of science saw that their chance had come. They had everything their own way. Praise of Soviet science poured forth in a stream. Public criticism of anything connected with the U.S.S.R. was made difficult, while praise could be given *ad libitum* (see p. 97).

The totalitarian idea of science has gained power as a result of the invasion of the U.S.S.R. by Germany. There are those who would have us believe that the success of the Soviet armies is evidence of the success of Soviet science. It is strange to reflect that there are people who say this, yet adhere to political groups which urgently opposed the rearmament of Britain. Six years ago they would have laughed, rightly, at the idea that military might is an index of scientific achievement. If military might were indeed such an index, then we should praise Nazi science above that of all other countries. German militarism is so efficient that nothing less than overwhelming odds in men and material will finally defeat it. An army is necessarily a dictatorship, not a democracy. As Walter Lippmann [67] explained before the war started, there

E

is one circumstance and one circumstance only in which totalitarianism is beneficial, and that one circumstance is war. If nearly everyone in a country has only one and the same intense desire, totalitarianism is the way to achieve it.   The defeat of an enemy in war is just such a universally desired end. In peace, however, people's ends are extremely diverse, and as Lippmann said, the achievement of diverse ends is best made possible by a liberal regime.   It is natural that totalitarian states which have long prepared for war should have greater military might than democracies which have not.   This throws no light, however, on the question whether science prospers best under a totalitarian or a liberal regime.

One must remember, further, that only a small fraction of the whole of science is applicable to the intentional killing of human beings and to their defence against being intentionally killed.   That is not the purpose of science.   Nevertheless, in the application of science to these ends, Britain certainly did not lag behind any other country when once she went to war.

3. *Science in the U.S.S.R.*

The Soviet government has made great progress in applying scientific knowledge to practical ends.   Russia was formerly backward in engineering.   By importing foreign engineers and training new Russian ones the government brought part of the country into line with Western Europe in this respect. Indeed, the rivers being larger, larger dams were constructed, and some of the more spectacular engineering achievements are well known.   That, however, is not what concerns us here.   Does a totalitarian regime provide an environment favourable to discovery?   Are intellectual standards in science high under such a regime?   Those are the questions with which I am concerned.

In the early days of Bolshevism, scientists were to a large extent left alone.   Looking through the scientific literature of the twenties, one finds some of the older Russian scientists still producing work of the same kind and quality as before the revolution. Central planning was so much occupied with the economic sphere that science escaped and to some extent prospered. The great physiologist, Pavlov, continued his research. First-rate work, like that of the physicist Landsberg, was produced. Soviet genetics prospered. Nawashin, who had been working at cytogenetics since 1912, continued his research. Vavilov, who began to be productive in 1915, continued his important work and did much to increase our knowledge of the origin of cultivated grains. New workers joined the field. Nassonov [75] in St Petersburg (later Leningrad) and Karpova [58] in Moscow made important contributions to our knowledge of the Golgi element in cells. Pavlovsky, who had published some studies on the anatomy and physiology of scorpions before the revolution,[78] continued to publish on the same subject after it.[79] Soviet scientists were not prevented from working as individuals. Good and bad work was done, as good and bad work is done in any other country.

Then totalitarianism descended upon science in the U.S.S.R. The subject had for the most part escaped from the first five-year plan, but it was involved in the second. The old Academy of Science was remoulded as a central planning authority for science. A five-year plan for science was produced, covering the period 1932–37. Seven subjects were selected for investigation, as follows [18]:

(1) the structure of matter, and its bearing on astronomy, physics, chemical physics, and chemistry;

(2) the survey and utilization of the natural resources of the U.S.S.R.;

(3) the survey and planning of the power resources of the U.S.S.R.;

(4) problems of distribution, building materials, hygiene, etc., arising out of construction;

(5) the general introduction of chemistry in industry and agriculture;

(6) the study of biological evolution, and the bearing of its results on agriculture and materials for light industry;

(7) the provision of the historical and social theory for combating the ideas of capitalism, and dissolving the prejudices which survive in the minds of the people, and have been transmitted from earlier forms of society.

It is a strange list. The first item is so wide that chemists and physicists could receive no direction from it. The second, third, fourth, and fifth items are technological. The sixth directs biologists into one field only that of evolution. It may be pretended that, since all organisms have evolved, all biological research is a study of evolution, but if so the item is so wide as to be meaningless. The seventh strikes curiously on the mind of a scientist in a democratic country, for instructions are given not only as to what shall be investigated, but as to what the results shall be. The list is a vivid illustration of the impossibility of planning scientific research. Either the subjects set must be so wide that it is futile to write them down, or else they must be so framed that they exclude the study of important branches of science. Both these tendencies are manifest in the list. Science can only progress rapidly if research workers are free to expand the bounds of knowledge wherever, at any given moment, they are expansible.

During the five years of the first plan for science, great strides forward were made in the outside world in fields from which Soviet scientists were excluded by the wording of the plan. Take almost any branch of non-evolutionary biological science in which outstanding discoveries were made in the outside world during the years of the plan, and you are likely to find that the whole subject was excluded from study. Sex hormones provide an example. The period of the five-year plan coincides with marvellous developments in that subject in Britain, the U.S.A., and elsewhere. The various hormones in their different forms were obtained as chemically pure substances and their structural formulæ determined. As a culminating feat of all this wonderful progress, œstrone itself was synthesized from cholesterol. Meanwhile other substances were synthesized from inorganic materials, which behaved like hormones when injected, and exerted œstrogenic effects. That whole subject was excluded from the five-year plan. "The survey and utilization of the natural resources of the U.S.S.R.; the survey and planning of the power resources of the U.S.S.R. ; the problems of distribution, . . . etc., arising out of construction"; and so on: there is no place here for work on sex hormones. It is strange to think of Soviet scientists being tied down to such matters as "the provision of the historical and social theory for combating the ideas of capitalism," when so much of enormous interest and importance in real science was being done in the outside world.

Genetics tells the same story as endocrinology. In this subject also marvellous discoveries were made in the outside world during the currency of the five-year plan. We already knew that the chromosomes bore the genes responsible for inheritance, but the chromosomes usually looked much the same all along their length and never showed enough irregularities of structure to account for all the genes which were known to exist. There was compelling indirect evidence that the genes were strung out along the chromosomes in a line, and we even knew the order of their arrangement. Here, at this point on a chromosome of the fly *Drosophila*, we could say, is the gene which expresses itself most obviously by its effect on the shape of the wings; here, further along the same chromosome, is another affecting the size of the legs; further again is a gene affecting body-colour, and further still one affecting the size of the wings; and so on for hundreds of other genes. But the chromosomes looked much the same all along, and naturally one wanted ocular demonstration.

That ocular demonstration has come, and in a totally unexpected way. It had been known since 1881 that certain large cells in the two-winged flies—cells of the salivary glands among others—have very extraordinary nuclei, with curious cross-banded tapes inside. It was not until 1933 that Heitz and Bauer [45] showed what those banded structures are. They are giant chromosomes, about one hundred times the length of ordinary chromosomes. With a little stretching we can make some of them not a few μ long, like ordinary chromosomes, but nearly half a millimetre. Now, it will be said, we shall see the genes. Heitz and Bauer showed that the cross-bands differ in thickness, and that they are always arranged in the same order. If, in a certain part of a certain chromosome, there is a thick band and next to it a thin one, and then two thicks and then three thins and a thick, that same arrangement occurs in all the cells of normal members of the species. We knew from genetic evidence that the genes are arranged in a definite order: now we could see bands in the chromosomes arranged in a definite order, and so numerous that they might actually be the genes.

It was Painter [77] and his co-workers in the University of Texas who made the crucial discovery. There are certain individual flies in which some of the genes are known from the results of breeding experiments to be, as it were, the wrong way round. If we call the genes of one chromosome of a normal fly *a*, *b*, *c*, *d*, *e*, *f*, etc., in alphabetical order, there are some abnormal flies in which a part of the chromosome, say *l*, *m*, *n*, is arranged the wrong way round, *n*, *m*, *l*. The complex evidence for that was wholly indirect. Now Painter set out to look at the giant chromosomes of these abnormal individuals. What he found is one of the big things in the history of biology. The order of the bands was reversed in the corresponding part of the chromosome concerned. The last link in the evidence was forged and final ocular proof given of the chromosome theory of heredity.

These wonderful discoveries in cytogenetics deserve comparison with some of the greatest of physical science. They are mentioned here for a very particular purpose. They belong to a part of biology omitted from the five-year plan covering the period when they were made.

People may perhaps be inclined to reply that research was in fact done in the U.S.S.R. on some of these subjects despite their exclusion from the five-year plan. This is irrelevant to the argument, for it is the central planning of research that I condemn, not disobedience to central planning.

4. *The Soviet Genetics Controversy*

Not only did the five-year plan omit those very kinds of study which gave such momentous results: omission was not all. Soviet authority lent its aid to those very people who wished to eradicate the whole subject of genetics. That was the intention of Academician T. D. Lysenko and his associates at the conference held in 1939 under the auspices of the journal *Pod Znamenem Marxisma* (*Under the Banner of Marxism*). [60] The proceedings at that conference deserve to be better known, for they throw a strong light on the state of science under a totalitarian regime.

The chromosome theory of inheritance, as we have seen, was proved by ocular demonstration by Painter and his associates. It was one of the greatest scientific discoveries of the century, and the crowning achievement of cytogenetics. Let us hear the Soviet scientists on the subject. Lysenko says that "the only thing left from the so-called Morgan chromosome theory of inheritance is the chromosomes, and the whole theory of Morganism collapses." This is a short way of dealing with the great American scientist, T. H. Morgan, to whom cytogenetics probably owes more than to any other single man. I. I. Prezent [85] had previously written that cytogenetics should be thrown "into the archives of delusions."

The Soviet biologists by no means stop short when they have dealt with the chromosomes; it is the whole of genetics that Lysenko and his associates would destroy. ". . . I do not acknowledge Mendelism," says Lysenko; ". . . I do not consider formal Mendelian-Morganist genetics as a science." He announces that he has made a public denial of the existence of the 3:1 ratio as a biological law. "It is possible to defend the false bases of Mendelism only by lies," he says; ". . . the teaching of Mendel and Morgan I cannot call anything but false." The great Danish geneticist, Johannsen, is selected for particular attack. "I quarrelled with Johannsen's theory," says Lysenko, "not because I dislike Johannsen himself, but because the Mendelians support his theory and make propaganda for him in our higher education courses."

V. K. Morozov rallies to Lysenko's cause. "The representatives of formal genetics say that they get good 3:1 ratio results with *Drosophila*. Their work with this object is very profitable to them, because the affair, as one might say, is irresponsible . . . if the flies die, they are not penalized." (This sentence is constructed sarcastically in a way which does not admit of a literal translation into English.) S. N. Davidenhov tells how famous doctors have advised him, "Stop bothering with genetics; the word heredity should not be mentioned."

A favourite way of attacking genetics was to say that it is metaphysical, though this charge cannot be substantiated by any valid argument. I. I. Prezent refers to Johannsen as "the classical representative of metaphysical genetics." "When he grasps Bolshevism," says Lysenko, "the reader will not be able to give his sympathy to metaphysics, and Mendelism definitely is pure, undisguised metaphysics." "The teaching of pure lines," says Professor L. N. Delonei, "is not free from metaphysics." The geneticist, Academician M. M. Zavadowsky, feels it necessary to answer the charge that the meaning of the words genotype and phenotype is metaphysical. Professor M. B. Chernoyarov concludes that "the chromosome theory of heredity is a metaphysical conception."

Openly political arguments are freely used for the purpose of attacking genetics, and Zavadowsky has to answer the accusation that "genetics inevitably, from its conceptions, leads to reactionary ideas, to the race theory." Professor N. P. Doubinin has to discuss not only whether Mendelism "exists," but whether it "represents a product of the imperialist development of capitalist society." He speaks of "a reactionary attempt to replace the philosophy of dialectical materialism." "Of course," he says, "after its appearance Mendelism was perverted by the bourgeois scientists." Govorov's point of view is described as "gravely inimical to dialectical materialism." Lysenko says that it is dialectical materialism that makes him deny the Mendelian law of segregation. "We must proceed from dialectical materialism," says Professor Polyakov; "it is from this position that we must appreciate what genetics has contributed and what Comrade Lysenko has to offer." "Soviet biologists," says M. B. Mitin, the chairman of the conference, ". . . must master dialectical and historical materialism, and learn to apply the dialectic method to their scientific work."

The charges that genetics is metaphysical and that it is contrary to dialectical materialism are by no means the only ways of attacking it. It can be ridiculed, and palpably false arguments can be used. "Three take after papa," laughs K. A. Timiryazev, "one after mama, or *vice versa*, three after mama, one after papa." (Timiryazev evidently thinks that the 3:1 ratio appears in F. 1.) Lysenko refers to the 3:1 ratio as "devil's work." Professor Polyakov tells the geneticist N. I. Vavilov, "it was not necessary for you to humble yourself before foreign science." L. N. Delonei says that Vavilov "raves about the Americans."

Arguments against genetics are put forward which stand self-condemned. "The whole Soviet people," says V. K. Milovanov, "thousands of specialists and collective farmers who are doing wonderful work under his supervision, are with Lysenko." The "small wilting group" of geneticists is "cut off from practical life." "In order to get a particular result, one must want to get exactly that result," says Lysenko; "if you want to get a particular result, you will get it." (These words were quoted by U. Y. Kerkiss, and Lysenko interrupted his remarks to confirm their correctness.)

The teaching of genetics was condemned by several speakers. "It is essential to remove the teaching of genetics from the secondary school," says Academician A. F. Yudin. "If young people are forced to study genetics, their ability to think will be damaged," remarks M. B. Chernoyarov. "To this day we have chairs of genetics," says V. K. Milovanov; "they should have been liquidated long ago."

The conference lasted seven days. Throughout, the word "Darwinism" was used in a special sense in which it is never used, so far as I know, outside the U.S.S.R. Apparently it means roughly what the outside world calls neo-Lamarckism, but it is difficult to be certain. Members of both camps— those who attacked and those who made a prudent defence of genetics — strove to show that they were genuine "Darwinists" and that their views accorded with dialectical materialism. The level of the discussion was very low. Scorn was thrown on the use of statistical methods. The opponents of genetics never got beyond the 3:1 ratio, except that in one place Lysenko mentioned the 1:2:1 ratio. It is as though a modern discussion were being held on the structure of the atom, and one party never got beyond the proportional combining weights of the elements and spent their time denying the "existence" of these proportions.

The party which attacked genetics, headed by Lysenko, is the one favoured by Soviet authority. This man is not only an Academician, but also the Director of the Academy of Agricultural Science. At his suggestion the Commissariat of Agriculture issued an order compelling workers at all agricultural research stations to carry out their work in accordance with Lysenko's belief in the inheritance of acquired characters (or, more precisely, in the absence of distinction between phenotype and genotype). The Soviet authorities also stopped the publication of translations of foreign books on genetics. These facts were recorded at the conference by N. I. Vavilov.

The report of the conference was reviewed in *Nature* by Mr P. G. 'Espinasse.[29] Anyone reading the review would conclude that for the most part the discussion was learned and concerned with highly technical problems. This is misleading. Mr 'Espinasse notices that argument does not get beyond the 3:1 level, but he passes this off by saying that the speakers used the term 3:1 ratio as standing for every ratio possible in genetic theory. A careful reading of the report leads me to reject Mr 'Espinasse's opinion, and to conclude that the arguments against genetics stood at a low level of scholarship because those who used the arguments had only an elementary acquaintance with the subject. Many English-speaking scientists probably have little knowledge of the conference and it is therefore desirable that the whole report should be published in English, so that British and American scientists may judge for themselves.*

Mr 'Espinasse allows that some of those who attacked genetics at the conference "made it quite clear that they did not understand what it was that was being discussed." Among the people who offered to produce desirable varieties of plants and animals, he says, were some who were "honest, ignorant, and useless" and others, "possibly to a considerable extent, consciously fraudulent." These few words do not suffice to present to readers a true picture of the low standard of scholarship exhibited at the conference.

The main lesson to be drawn from the Soviet genetics controversy is that science can flourish only if free from state control. Professor M. Polanyi has particularly emphasized this point in his article on "The Autonomy of Science," [84] which deserves to be read by every scientist. If the selection of scientific personnel is left to the state, the wrong men are likely to be given important posts, because those who are not

---

* A translation has been made for the Society for Cultural Relations with the U.S.S.R., which has been available to me. The translation is useful, but it has not been made by a geneticist and is not very literal in places. The School of Slavonic and East European Studies has kindly made the original Russian version available to me.

themselves, scientists will be led astray by the false claims and pretences of ignorant and foolish persons. As we have seen, such persons may even become academicians and be given wide powers to control scientific research. Worse still, scientists may exhibit a servile obedience to their political bosses and let dogmas and slogans affect their science.

5. *The Deficiencies of Soviet Science*

The impression that Soviet science is great tends to be fostered by continual repetitions of praise. The same discoveries are praised repeatedly, while greater discoveries made in non-communist countries go without mention. One could take almost any European country and find things to praise in the scientific work done there, as well as excuses for shortcomings. If one went to that country, visited various laboratories and chatted with the research workers, one could come back and write an enthusiastic account of what one had seen and heard.

No one will deny that good scientific work has been done in the U.S.S.R. since the inauguration of the five-year plans for science. Excellent work has been done in ecology, for instance, by D. N. Kashkarov, N. I. Kalabukhov, G. F. Gauze, and others. My friend Mr Charles Elton tells me that Soviet ecological research has been on the whole similar to that of Britain and the U.S.A., except that the Soviet ecologists have very poor statistical training and judgment. Other subjects in which good research has been done will occur to specialists in various fields, and some of them have been so much advertised that they are generally known. Many countries, however, produce good science, and it would be false to pretend that Soviet scientific research as a whole would bear comparison with British or American, whether one looks especially for particular excellence or absence of particular weakness. One may argue, of course, that the particular circumstances of the U.S.S.R. after the revolution made it necessary to concentrate on applications and to regard research in science as of less importance. If this argument be accepted, one should still withhold praise where praise is not due.

Even the most enthusiastic writers about Soviet science have to admit its deficiencies. Sometimes it is done very mildly. Dr J. Needham tells us that when he visited Russian laboratories in 1935 "the general standard of work was not quite up to the prevailing standards in Western countries, or in the United States," but he hastens to prophesy rapid improvement.[76] Dr D. Shoenberg allows that "there seemed to be a good deal of unnecessary red tape" in the Institute of Physical Problems in Moscow, where he worked in 1937–38; and he remarks on the difficulty of getting things for use in scientific work, which one could buy at Woolworth's in Britain. He complains also of "a rather too literal interpretation of planning, in the sense of too slavish a copying of the type of plan suitable for industrial production." He allows that "there have as yet been no really revolutionary developments of importance comparable with Rutherford's work on the atom, or advances in the technique of physics of importance comparable with Lawrence's invention of the cyclotron." He writes also of the uncritical attitude of Soviet physicists and their "over-easy acceptance" of their own conclusions.[92] Dr A. Walton writes that much of the research in the U.S.S.R. was "at first somewhat immature and even primitive."[107] Mr J. G. Crowther[19] says that the young scientists are still "immature." Some Soviet scientists, he admits, denounced scientific theories because they were advocated by persons whose political views were regarded as inimical to the state. The scientific researches of persons disapproved by the political authorities have sometimes been omitted in lists of references, he says, and scientists have sometimes been seen apologizing to the political authorities for having held opinions which appear to the majority of the scientists in the world to be correct. Many scientists have been imprisoned and some shot. Elsewhere[20] Mr Crowther says that the Soviet system has produced "tens of thousands of new young scientific workers, many of them somewhat raw and not yet old enough in tradition to have highly developed scientific intuitions and critical judgments, but superb material for flying and artillery officers and factory management." Professor J. D. Bernal [10] says that it would be "idle" to look for "quietly pursued excellence and sound and acute scholarship" in the U.S.S.R. Elsewhere [8] he writes of the "shortcomings and backwardness of Soviet science," and elsewhere again [9] of "the errors and the crudities of Soviet science."

These quotations are from those who write in *praise* of Soviet science. It is legitimate to ask whether the same authors would have written in praise of something so admittedly open to adverse criticism, if it had been the product of a democratic state.

Those who praise Soviet science are not ashamed to use directly contradictory arguments. On the one hand, they extol the virtues of the central planning of research, and pour scorn on the individualist who wants to choose his own subject. They ridicule the study of pure science, and say contemptuously that they do not recognize a part of science as "impure." All this, of course, is the central theme of those who praise Soviet science. The same people, however, use arguments which flatly contradict their main contention.

Mr J. G. Crowther, in his *Soviet Science* (Kegan Paul, 1936), [18] tells us on pp. 192–193 that in England "the majority of chemists are too much restricted to immediate problems . . . the chemist in the Soviet Union is much more free to choose the sort of problem he would like to investigate. If he wishes to undertake fundamental research, he is encouraged. In England there is far too much control by authority: the chemist is too frequently expected to attack problems chosen for him by his superiors." These statements are made on the authority of a British scientist who worked in the U.S.S.R. On pp. 87–88 Mr Crowther tells us exactly the opposite, in discussing "The most original feature of Soviet laboratory organization," the central planning of research. This he illustrates by the example of the Kharkov Physico-Technical Institute. He explains that the plans for research, drawn up by a central authority, "must receive adherence in principle from every research worker. . . . They are guides which prevent any worker from being uncertain what problem he should be attacking. The research worker cannot change the subject of his researches without wide discussions by the staff of the institute. . . . The personal desires of the individual members of a brigade receive little consideration . . . the Soviet research worker does not count much as an individual." The central planning organization for the whole of Soviet science is the Academy, whose first duty is "to plan and direct the study and application of science towards the fulfilment of socialist construction, and the further growth of the socialist order" (p. 24). Now which of these stories is true? Is the research worker free to choose what he will study and is he encouraged to undertake fundamental research, as stated on pp. 192–193, or is the exact contrary true, as stated on pp. 87–88?

There has been much propaganda for Lysenko's methods of controlling the time of germination of seeds, and people might be led to think that "vernalization" was a Soviet discovery. An American scientist tells us, however, that the discovery was made in the United States before the Civil War.[91a] The principle was certainly firmly established by Braem in 1890, in connexion with the germination of the resting reproductive bodies of animals.[11a]

Every country produces poor as well as good science, but the U.S.S.R. has produced an unduly high proportion of bad and suspect science, such as the work on the artificial control of the sex ratio and on "mitogenetic rays." It is fair, however, to add that the existence of these supposed rays has been denied by certain Soviet workers as well as by the scientists of the outside world.

It is difficult to assess the scientific achievement of one

country in comparison with others, but the following method was devised for the purpose of getting some unbiased information on the subject. I asked seven scientists to prepare a list of the two dozen most important scientific discoveries made between the two great wars. I gave them no hint of the reason for my request. They are all lecturers in science at Oxford, who themselves carry out scientific research in the University Departments of Physical Chemistry, Organic Chemistry, Botany, Zoology, and Physiology. They provided me with a list of twenty-seven items. The list is as follows. (Three of the items are inventions rather than discoveries.)

Schrödinger, Heisenberg and Dirac on quantum mechanics.

Rutherford on the structure of the atomic nucleus.

Aston on isotopes.

The Braggs, Goldschmidt, Debye, Bernal and Astbury on molecular structure as exhibited by X-ray diffraction and other physical methods.

Langmuir on molecular forces as exhibited by the study of surface films.

Curie-Joliot on artificial radio-activity.

Lawrence on the cyclotron.

Bush on the electron-microscope.

Svedberg on the ultra-centrifuge.

Sidgwick and Pauling on molecular structure and valency.

Carothers on high polymers.

Haworth on the molecular structure of carbohydrates.

Karrer and Kuhn on carotenoids.

Szent-Györgyi and Williams on the isolation and synthesis of vitamins.

Northrop on crystalline enzymes.

Freundlich on colloid chemistry.

Stanley on virus diseases.

Went on plant hormones.

Fisher, Haldane, Timoféëff-Ressovsky, Muller and Darlington on the genetical theory of evolution and allied subjects.

Spemann on chemical factors in animal development.
Vogt on cell-movements in animal development.
Dale, Loewi and Feldberg on the chemistry of nerve action.
Landsteiner on serological differences between individuals.
Domagk on sulphonamides.
Fleming and Florey on penicillin.
Cook and Kennaway on carcinogens.
Fisher on the statistical treatment of small samples.

Different groups of scientists would have produced different lists, but a good proportion of the items are, I think, almost inevitable. The striking fact about the list is that it contains no mention of Soviet science. So far as I know, there has been little or no propaganda for British, American, or German science, and it is interesting to find that all the propaganda for Soviet science has not resulted in the inclusion of a single item from the U.S.S.R. Muller, it is true, did part of his work in that country, but he is an American who made his fundamental discoveries in the University of Texas before he went to Russia. Timoféëff-Ressovsky did some of his early work at the Institute of Experimental Biology in Moscow, but the research for which he is famous was done at the Kaiser Wilhelm Institute in Berlin. The list suggests what is likely to be the general opinion among scientists, namely, that Britain, the U.S.A. and Germany were pre-eminent in scientific research in the period between the two wars; though there is strong evidence that totalitarianism was beginning to have a detrimental effect on German science in the thirties.[36a]

It might be thought that I chose seven scientists who were politically biased against the Soviet Union and were therefore unable to make a fair selection (despite the fact that they did not know the purpose of the list until it was made). For this reason I mention that I chose two socialists among the seven. Of the others, two adhere to no party, while the political views of the remaining three are unknown to me.

Those who praise Soviet science are sometimes almost

F

pitifully anxious to make the most of small discoveries.   Thus the distinguished Soviet scientist, Dr Peter Kapitsa,[57] in a general survey of science and war in the U.S.S.R., tells us that Soviet scientists are experimenting with a synthetic drug which is likely to have curative properties "not inferior to those of Peruvian balsam." This is his only claim for Soviet pharmacology.   Balsam of Peru is still sometimes used in the treatment of wounds.   It was introduced into European medicine by Nicholas Monardes of Seville in 1560.[34] Monardes would have been flattered if he could have realized that nearly four centuries later the preparation of a "not inferior" substitute for his balsam would be used in propaganda on behalf of a large country.   Meanwhile scientists in Britain were at work on penicillin.   There has been too much boosting of discoveries at the level of a substitute for balsam of Peru.

6. *Party Politics and Free Science*

The movement for free science and pure science is not political in any narrow sense.   It includes liberals, conservatives, and socialists.   That is as it should be.   It goes without saying that liberals favour free science.   Conservatives want to conserve what they think good in existing institutions, and those of them who think free science good naturally want to conserve it.   There is, I think, no essential element in socialism opposed to the ideas that pure science has value as an end in itself, that technology must have science as its basis, and that science prospers best when research workers are free.   It is only from believers in totalitarianism that we can expect no support.   The central planning of science is essentially part of the totalitarian theory of the state.

Chapter V

THE DUTIES OF SCIENTISTS TO SOCIETY

"The philosopher * should be a man willing to listen to every suggestion, but determined to judge for himself. He should not be biased by appearances; have no favourite hypothesis; be of no school; and in doctrine have no master. He should not be a respecter of persons, but of things. Truth should be his primary object."— MICHAEL FARADAY, scientist.[31]

"Understand, we are not here to pass honest judgments but to purvey matter the other side doesn't like."—DAVID LOW, political cartoonist.[68]

1. *Introduction*

SCIENCE is necessarily a social, activity. It is true that the research worker is often rather a retiring person, finding little pleasure in the kinds of social life that mean so much to others. Nevertheless, that his work is social is shown by the necessity for demonstrable proofs. This necessity presupposes the interest and participation of others. The person who makes a systematic collection of natural objects simply for his own satisfaction, without adding to the common store of knowledge or infusing others with his interests, is not a scientist. It is true that the scientist's motives may be and often are mixed. Unless he has an intense internal urge to find out, to try to satisfy an insatiable curiosity, he will not be much of a research worker. Still, his work must be presented in a form in which it will influence the opinion of others.

Although no one should be forced to disclose his results before he has satisfied himself that they are fit for publication, yet there ought to be no private property in demonstrable knowledge. If a scientist keeps his discoveries to himself, his activities are unsocial and indefensible; for knowledge is good, and it is better for many than for one person to possess it. Every scientist lives a vastly fuller life just because generations of other scientists have not kept their discoveries to

* Faraday's word for scientist.

themselves. To profit from the work of others without making the return that is within one's power is indefensible.

Newton has often been charged with secretiveness, but perhaps his case has been rather misrepresented. It is true that he did not hurry to get his discoveries printed, but this was largely because of his great dislike of controversy. He was elected a Fellow of the Royal Society at the age of twenty-nine. Five days before his election he wrote to the Secretary saying that if elected "I shall endeavour to testify my gratitude by communicating what my poor and solitary endeavours can effect towards the promoting your philosophical designs." [95] He was by no means slow in keeping his promise. Exactly a month later he sent to the Secretary his famous paper on the composition of white light, which he showed to be made up of a spectrum of different colours, which were refracted by glass to different degrees. It was the reception of this paper that turned him in upon himself and made him so secretive. Professor Linus of Liége, failing to understand Newton's experiments, criticized them stupidly in a paper published in the *Philosophical Transactions* of the Society. Unfortunately, the Secretary pressed Newton to reply. The matter dragged on for years, Newton refusing to answer because he thought that Linus had not taken the trouble to understand what was clearly explained in the original paper. Finally Newton became exasperated. "I see I have made myself a slave to philosophy," he wrote to the Secretary, "but if I get free of Mr Linus's business I will resolutely bid adieu to it eternally, excepting what I do for my private satisfaction, or leave to come out after me; for I see a man must either resolve to put out nothing new, or to become a slave to defend it." Newton did not stick to his threat, but this incident certainly made him more secretive; and science would have profited even more generously than it did from his genius if he had been readier to publish his discoveries.

The primary duty of the research scientist, as such, is to make the greatest possible public contribution to demonstrable knowledge. He generally requires no advice from anyone on the subject of his primary duty, for he knows better than others how his own particular qualities can be used most effectively. There are some pitfalls, however, which may be regarded as of a moral kind, even in the accomplishment of his primary duty. He must be on his guard against lazily drifting into trivial work, limited in significance and unlikely to open new fields of investigation or help others in their research. Although he must make full use of working hypotheses, yet he must avoid the fascination of idle speculations. He must never depart from truth for convenience, as when a systematic biologist leaves unchanged a classification of organisms which he knows to be unnatural, simply because a particular grouping is usual or convenient. He must avoid the temptation to follow a course simply because it is fashionable, or likely to bring him praise because it is immediately connected with practical affairs, when he knows that his own particular talents enable him to discharge his primary duty better in another way.

The discharge of this duty is something in which the investigator finds pleasure. It may therefore be asked whether any extra or secondary duties are owed by the scientist to the community in return for being allowed to do what he likes doing; for many people have uncongenial employments which they would give up at once if they were made financially independent.

It must be remembered that the research scientist has often obtained his position in society by voluntarily giving up highly prized pleasures in order to devote himself to laborious study. No scientist is likely to get far whose spare time is all spent in frivolous amusement, or whose study is always devoted to those subjects that are immediately attractive to him. He knows that he must master certain difficult subjects, although he finds them unattractive, in order that he may understand better the subject of his choice. He voluntarily disciplines himself to study while others seek ephemeral amusement, and devotes himself to hard and uncongenial tasks. This discipline he imposed on himself when young, even while still a boy; if he had not done so, he would probably not be a research scientist. The recurring disappointments of research and the unavoidable tedium of certain parts of it must also be remembered. The community, then, should not thoughtlessly demand extra services of the scientist. Nevertheless, he often asks himself what services he can render to the community as a whole, apart from his primary duty; and the general congeniality and liberty of his task under the conditions of free science make him ask himself this the more insistently. The teacher of science, whose primary duty is to instil the spirit of science and to convey factual knowledge, will ask himself the same question.

When confronted with this ethical problem, the scientist may reach either of two conclusions. On the one hand, he may decide that his best contribution to human welfare may be made privately in the circle of his own acquaintance. He may believe that social evils will continue as long as people rely on politics and propaganda for human betterment, and that the only hope for progress lies in the voluntary improvement of the relations between those who come into direct contact in the ordinary course of their lives. He may have been impressed by the low moral standards of many politicians and propagandists, and by the solid good done inconspicuously by good men and women. On the other hand, he may consider that his special qualities as a scientist give him the privilege of a wider and special scope for good.

It is for the scientist himself to judge between the two courses. His temperament must be the deciding factor. If he chooses to limit himself to private action the decision is perfectly proper, and it would be an impertinence to deny him the right to make it. In that case there is no advice that a scientist can offer. If, however, he decides that he will use his special qualifications directly for the public good, apart from his primary duty, then certain observations may be offered, which will occupy the rest of this chapter. These observations are intended to indicate some of the accessory

ways in which the scientist can serve his fellow-creatures.
Anyone who makes suggestions on this subject lays himself
open to the charge of being an unqualified moralist, and the
task is not undertaken without diffidence. Nevertheless, it
is urgent that some one should undertake it, and the very
defects in what I shall write may prompt better-qualified people
to improve on my suggestions. Those who think that science
should be centrally planned are undeterred by doubts as to
their qualifications to give advice on ethical subjects. They
have spread powerful propaganda intended to make scientists
believe that they have three social obligations: to devote all
their energies to the solution of the problems of man's material
wants, to accept central planning in their own subject, and to
press for the adoption of central planning of society as a whole.
This, so far as one can make out, is what is meant by the social
responsibilities of scientists, as the term is usually used.
Scientists who have other ideas about their social responsibilities have not yet bothered to challenge the propaganda,
which is effective because it is not answered.

2. *Science and Evil*

The scientist is often charged with being responsible for
much of the misery and unhappiness in the world, because his
discoveries help the engineer to devise and construct the
weapons of modern war. When people reflect on human
miseries, they are apt to think first of those of war, and it
would be futile to seek to minimize them; yet it is doubtful
whether they are the most intense. The person bereaved,
wounded, or dying in war has the moral support, the fellow-
feeling of his compatriots. The sufferer from incurable
disease, again, has at least the relief that comes from the
knowledge that friends, doctor, nurses—all with whom he is
in contact—strive hard to lessen the suffering. Far more
intense is the misery of those who are persecuted by authority
for racial or political reasons. The concentration or penal

camp, the activities of the secret police (whether it be Gestapo, Ochrana, or Ogpu), the totalitarian purge—these are surely the greatest evils of mankind, by which individual human beings, because their beliefs and ideas differ from those of a central planning authority, are taken away from everyone whom they hold dear, kept under conditions of appalling cruelty, repeatedly subjected to questionings and to threats directed at relatives, and finally, in many cases, either executed after a so-called trial, shot in prison without even that, or killed through exposure to intolerable conditions.[59, 69, 99, 100] These evils are the greater because of the huge number of people who have suffered them. They are probably the greatest evils known to man, and no part of them can be ascribed to the discoveries of science.

Man's deliberate inhumanity to man, which is probably the chief cause of intense suffering, is something for which the scientist is not responsible; though as knowledge of psychology grows he may become able to lessen it. Ruthlessness is a characteristic of totalitarian but not of democratic states, and psychologists may one day be able to prevent ruthless cruelty by discovering the causes that make people accept totalitarian government. Again, cruelty is often the result of hate engendered in childhood, and a better understanding of child psychology may reduce the amount of hate in the world.

Scarcely anyone will deny that modern war is a terrible evil, and there are those who would charge scientists with the responsibility. It is suggested that an urgent duty of scientists is to make sure that their discoveries are not used for destructive purposes, especially in war. This, however, would make scientific research impossible. Professor P. W. Bridgman, [12] of Harvard University, has stressed this point. If, as he says, the scientist were required to make only those discoveries which could not wilfully be perverted to harmful uses, he would almost certainly feel himself so restricted that he would make no discoveries at all. It is impossible to foresee all the consequences or to balance the good consequences against the bad. There is no mechanism by which the scientist can control the consequences of his discoveries. As Professor Bridgman says, it is society as a whole that is in a position to provide the mechanism of control rather than the individual discoverer.

Again, the fact that greater numbers of people are killed in modern than in ancient wars is due not so much to the greater effectiveness of modern offensive weapons (for defensive devices are evolved to meet them), as to modern methods of sanitation and transport, which enable vast numbers of people to congregate together in cities or armies and to be killed in huge numbers. Even if technologists did not invent explosives, and soldiers relied on clubs and swords, huge numbers of people would nevertheless be killed in modern war. Nearly everyone, incidentally, would rather be wounded or killed by a bullet or shell than by a bayonet or sword. The person who condemns science because of its relation to war must be prepared to abandon modern sanitation and transport, the chief causes of huge casualties.

3. *Some Secondary Duties*

So much for the negative aspect of the problem. We have discussed the evils for which the scientist cannot be held responsible. What good can he foster and what evils can he prevent?

First comes the preservation of our scientific heritage. Chapter II was devoted to the values of science. We have seen that knowledge has been built up by people who valued it as an end. Some of our younger scientists, affected by propaganda, are ready to let central planning and a crudely materialistic outlook supplant the freedom and idealism that have made science great. Our scientific heritage is threatened. Let us understand that science is not our private fortune, to squander as we will. It is a heritage that is entailed to future generations, and we should preserve it and add to it as much as we can before we pass it on.

Science is not simply a mass of demonstrable knowledge about nature recorded in books and journals. That is not in danger, for there is no threat to burn our scientific records. The part of our heritage that is threatened is the spirit of science. Little has been written about that spirit. As Professor Michael Polanyi [84] has remarked, it is something that cannot easily be communicated except by personal example. It is only a few centuries old, apart from vague beginnings such as we see among present-day savages and sporadic appearances in concrete form throughout the historical period. There is no certainty that it is immortal. Only ignorant and thoughtless people can take it for granted, or imagine that they would have possessed much of it if they had not received it from others. Here is a field in which research workers, as well as students and teachers of science in schools and universities, can render a great service to the community. We can all work to maintain and enlarge the belief in the value of science as an end.

We can do much to preserve our scientific heritage. We can dispute openly with those who profess to see nothing in science except an easier supply of food, shelter, health, and leisure. As we saw in Chapter II, it is fantastic to suppose that while music and art are ends, science is not. We should urge young scientists with real aptitude for science itself to resist the higher salaries that will be held out to tempt them into technology. We should do everything in our power to make as many people as possible understand that science is an end.

That brings us to our second social responsibility. If science is good, then the greater the number of people who enjoy it, the better. There is scope for the encouragement of amateur by professional science, a subject to which I have devoted a chapter of another book.[6] The amateur, by definition, is one who values science as an end and thus possesses the scientific spirit. In the past the contributions of amateurs to scientific knowledge have been enormous, and even in modern times excellent research is carried out by amateurs, especially in biology and geology.[6, 103] There are amateur scientific societies of high standing, such as the Quekett Microscopical Club, and the Malacological and Conchological Societies. Amateur science presents a field in which suitable professionals can help both amateurs and themselves, to the advantage of science; though it must be admitted that certain pitfalls present themselves. If the professional who finds happiness in the direction of others co-operates with amateurs merely by collecting round him a group of unpaid helpers who like being told what to do, there may be satisfaction of various psychological urges on both sides, but little progress in science is likely to come of it. The professional who wants to co-operate genuinely in this movement must put aside feelings of authority and condescension, and be as ready to learn as to teach and direct.

It is unfortunate that love of science scarcely exists to-day among the poorer classes of society. It was not always so. One thinks at once of Hugh Miller, the Scottish quarryman who observed the rocks and went on to become the author of *The Old Red Sandstone*, a classic now reproduced in Everyman's Library.[73] Miller was not unique. Parallel to his life-story runs that of a man immortalized in the name of *Peachia*, a remarkable genus of sea-anemones. Charles Peach was a very exceptional man. He had four shillings a day, a wife, and nine children. He was only a private in the mounted guard (preventive service) at an obscure part of the Cornish coast.[55] I have not been able to discover exactly what it was that he was paid four shillings a day to prevent, but I do know that in his spare time he was a first-rate marine naturalist. Annually he attended the meetings of the British Association and mixed on level terms with the best contemporary biologists. Men of his calibre seem to have become nearly extinct. Peach shows that poverty is no bar to the study of science, but as wealth becomes more evenly distributed it ought to become increasingly easy to spread a love of science through all sections of the community.

The third social responsibility of scientists follows naturally on the second. They should take steps to facilitate the entry of suitable people of all classes into the profession of science. It must never be forgotten that genius and talent have sprung from every class, and that it is wasteful to let any potential talent remain unrecognized as a result of lack of opportunity. The scholarship system will bring a lucky and precocious child to the university without cost to the parents, but that is not enough. We need some comprehensive arrangement for bringing suitable children—and men and women of all ages too—into personal contact with first-rate research workers. Nearly every successful scientist can recollect some early incident in which his enthusiasm was first kindled. Many potential scientists in all classes of society probably go through life without ever experiencing such a kindling of enthusiasm. Something could be done if those who owe their position and success in science to such an incident were to give the same kind of help to others. It is doubtful whether popular books, radio talks, and public lectures fill the need (though it was a public lecture that started Benjamin Franklin on the course of action which made him become one of the world's foremost physicists through spare-time work). For adults the W.E.A. is, of course, in many ways excellent, but it probably serves to transmit factual knowledge rather than the spirit of science. The transmission of that spirit can usually be best effected by informal personal contact between the potential scientist and some one in whom intense enthusiasm has already been awakened. Here is a field in which scientists who have the necessary human understanding can quietly and unostentatiously do good not only to those whom they inspire, but also indirectly to the whole community.

Right through the history of science, people in a position to do so have noticed and encouraged talent. One has only to think, for instance, of what the Duke of Brunswick did for the bricklayer's son, Johann Gauss, and through him for science and humanity. As we move towards a healthier society in which great wealth and poverty will no longer co-exist, so the kind of help necessary will change. It is stil' true to-day that the child born in the richer classes has a great advantage, for it is generally much easier for him to get into personal contact with first-rate research workers. This is wrong and wasteful, and scientists have it in their power to correct the error.

Beyond those who could be inspired to great heights in this way, there are others who have the spirit of science but not the kind of mind that makes for academic success. There is scope for another entrance to the world of science than that which leads through the examination-room. It might be profitable to reform altogether the status of the laboratory assistant. At present assistants tend to be recruited in a haphazard way. The senior assistant often suggests to a parent that a boy should join the laboratory staff, without anyone considering whether the boy himself has any real feeling for science. Such boys sometimes fail to get that feeling, while others develop it gradually and not only become first-rate in their work, but have the satisfaction of living full and worth-while lives. What is wanted is some scheme whereby people can become laboratory assistants because they want to be scientists. This would be linked to a scheme enabling them to pass on eventually, if suitable, to independent research. Famous scientists like Faraday and Rühmkorff have forced their way into the world of science by beginning as laboratory assistants, and the process still happens to-day; but science would be well served if this entry into its ranks were made less fortuitous. The whole status of the laboratory assistant could be improved, and people of all classes encouraged to enter the scientific world as assistants because they felt that their real life-interest lay in that subject.

4. *Science and Politics*

The fourth social responsibility of scientists is in relation to politics.

The scientist knows how small the uncertainties are in his own sphere compared with the uncertainties of political doctrines; yet he knows that the scientific outlook changes radically as a result of new discoveries. This teaches him not to be a dogmatic adherent to any political party, and it also teaches the much profounder truth, that the irreversible must above all be avoided. The scientist knows that again and again he and other scientists are wrong in the conclusions they draw from factual evidence. He knows also that it does not matter, because nothing irreversible has happened as a result of his wrong conclusions. He is always ready for change. In politics, therefore, where everything is much more uncertain, he must raise his voice against all irreversible decisions, or decisions reversible only by bloody revolution. He is willing that any form of government whatever should be tried, provided that it can easily be reversed if people find, on free and open discussion, that they do not like it. For this reason it is consistent with the scientific outlook that he should oppose all tyrannical monarchies, such as that of the Czars, and all totalitarian regimes, whether national socialist, fascist, or communist. He may be a liberal, a conservative, or a socialist in his political views, or may hold any opinions whatever on economic matters; but if he is determined that, once his policy is put into practice, nothing but a warlike revolution shall change it, then he has left the scientific spirit behind him in the laboratory.

In order that social conditions may be continually improved by change, two principles of sound politics must above all be defended, and they are both principles upon which the scientist puts the highest value in his own sphere. They are the principles of free speech and valid argument. The first existed in peace-time Britain: the second has never flourished but is a necessity for real progress.

Not only has freedom of speech and publication failed to penetrate into powerful countries, such as Russia, where it never existed, and been ousted from others, such as Germany, where it seemed reasonably secure, but even in Britain there is a danger that it may not be restored after the war. We hear to-day the argument that economic liberty is more important than liberty of speech, an argument whose logical unsoundness does not prevent it from having a certain vogue. Even scientists, who see as clearly as anyone else that this freedom is of paramount importance if truth is to be attained, are willing to be equivocal on the subject where politics are concerned. Professor L. Hogben, for instance, writes thus: "If the functions of democratic government were still as when Milton claimed the right to know, to utter and to argue freely according to conscience, they would be equally important to-day. They are not." [46] Although it seems uncertain what the word "they" refers to in these sentences and the exact meaning is therefore not clear, yet this passage must apparently be taken to be a statement by a scientist directed against the necessity for freedom of speech in politics. Scientists would do well to register dissent from such views. In time of war, there has to be some limitation of freedom of speech and publication, but that is precisely because it is essential during war that the truth about secret military matters should not be known: thus the exception proves the rule. Whoever argues against freedom of speech does not wish the truth to be known.

There is one serious aspect of the subject of freedom of speech that is commonly overlooked. It is the support given to the cause of free speech by those who only want to use it to gain power and thus be in a position to eradicate it. That free speech is desirable is a liberal idea, directly contrary to communist theory and practice; yet communists support both the National Council for Civil Liberties and The Radio Freedom

League. That real believers in free speech should collaborate with those who wish to eradicate it is deplorable. Nothing would be easier than to prevent the adherence of dissimulators to causes in which they do not believe. Societies concerned with freedom of speech could require a signed statement from adherents that they believe in freedom of speech as a permanent safeguard of progress, and not as a temporary expedient to be cast aside when it has served the purposes of a party.

No freedom is so important as freedom of speech, for if that is free, other curtailments of liberty can be made known and redressed. With free speech, no period can be utterly stagnant, even when an unprogressive party is in power; for the circulation of new ideas will never stop so long as speech and publication are free.

The censorship on publication about the U.S.S.R. in the periodical Press, and the general acquiescence in that censorship, are among the most disturbing features of contemporary Britain. There is more than acquiescence. Even a formerly liberal paper, the *News Chronicle*, protests indignantly at the least criticism of the Soviet regime, when one would expect it on the contrary to be demanding the fullest right of every citizen to express his opinion. The method by which the censorship is exerted has not been fully explained to the public, but of its existence there is no doubt. For instance, the editor of the scientific journal *Endeavour* asked me to contribute a letter to "start a hare." I accordingly wrote a letter, urging the necessity for freedom in science and suggesting that the success in war of the totalitarian states, Germany and the U.S.S.R., should not make us think of totalitarianism as a progressive force in any other sphere. The letter was put into type and a proof sent to me. However, when the letter appeared in print, I found that the editor had rewritten part of it without my permission, so as to exclude the reference to the U.S.S.R. as a totalitarian state.[7] He had substituted other words of his own, not conveying the same message. When I protested, he answered that "we were assured in quite unambiguous terms that the original wording would not get by the censorship, and so had to bow to necessity." The editor's letter, in which these words occur, is open to inspection by any interested person. Meanwhile, alongside my letter appeared letters from two other scientists, both containing favourable comment on the Soviet Union. Newspapers of nearly every shade of political opinion pour out praise for the U.S.S.R., but scarcely ever print even a mild adverse criticism of any aspect of the regime.* The same applies to the B.B.C. programmes: constant praise is allowed, never (so far as my listening experience goes) a single adverse word since the U.S.S.R. was forced into war.

When Germany invaded the U.S.S.R., that country and Britain were in a position which has points of resemblance to that of two strangers who are attacked by a maniac in the street. The proper thing is obviously to collaborate in every possible way to restrain him. It is no time for back-chat between the strangers as to details of their home lives. Far the best thing is for each person to help the other as much as he can. That would have been a desirable and healthy relationship between the two countries. Directly the U.S.S.R. was invaded, however, those who favour its political system saw that their chance had come, and propaganda began. Although British socialists believe in free elections and parliamentary government and many other institutions unknown in the U.S.S.R., yet on this occasion they did not resist the opportunity of praising a state which had put into practice certain of their own precepts. The public, which had been strongly anti-Soviet at the time of the war against Finland and when the U.S.S.R. made a pact with Germany, now began to change its mind. As more and more pro-Soviet propaganda poured forth and no answer was published * (except in obscure journals written in foreign languages), public opinion underwent a profound change. People began to believe that those who

* The *Spectator* is an exception to this statement. It has boldly upheld the principle of freedom of publication, so far as the censorship allows.

G

had criticized the U.S.S.R. were dishonest people who had given false information for their own private benefit. They now became anxious to receive more and more pro-Soviet propaganda, and were incensed against even a very feeble voice raised in protest. Those who had kept quiet about the virtues of the Soviet regime during the episodes of the Finnish war and Stalin's pact with Hitler now became vocal, and to the voice of genuine believers in totalitarianism were added those of demagogues who are always ready to tell the public what it wants to hear. Among those who praise the Soviet regime most loudly are people who only a few years ago attacked it in words which would be quite unprintable to-day*: and they give no public indication that they have changed their minds. The propaganda has increased in snowball fashion, the public demanding more and more praise and resenting ever more fiercely the slightest dissentient voice.

No scientist, except one who keeps his mind divided into watertight compartments and uses scientific method only in the laboratory, can view the present situation with equanimity. One side only is being heard, and every scientist knows that the truth cannot be attained in that way. The public acquiescence is a danger-signal. We need another Milton, a Hazlitt, a Bentham, or a Mill to remind us that Britain was formerly pre-eminent in the preservation of liberty of speech and publication. Nowadays those who love liberty have got to admit that it is the U.S.A., not Britain, that upholds their cause. It is unlikely that anything will change the government's decision on this matter while the war lasts, but scientists have a solemn duty to demand the restoration of freedom of speech and publication directly it ends.

Another political duty of scientists concerns the principles of valid argument. In a democracy, legislation depends ultimately on the opinion of the people as a whole. It is therefore desirable that they should form their opinions by considering valid arguments based on a truthful presentation

---

* I am ready to substantiate this statement directly the censorship is lifted.

of facts. It is my purpose to suggest that the scientist's love of truth could be harnessed to the common welfare of mankind, if he could persuade others to use his own method of argument in the political sphere.

Three kinds of people above all others are concerned with the fearless discovery and dissemination of truth. These are the philosopher, the historian, and the scientist. All three, if they wished, could use their influence to improve the standard of political discussion. The scientist, however, is likely to be able to influence the mass of humanity more easily than the philosopher or the historian, because people tend to suspect philosophy and ignore history, while trusting and respecting science. This is perhaps not very extraordinary. The differences of opinion between philosophers have always been more fundamental than those between scientists. Philosophers have not built up a body of generally accepted opinion at all comparable in magnitude with that built up by scientists. Although people do not often dispute with the historian, yet they do not see his actual evidence; and as a rule they are far more interested in the present and future than in the past. These remarks are not intended as a disparagement of philosophy and history, than which scarcely anything could be more absurd: they are only intended to suggest that the scientist can probably influence the opinion of the people more easily than the philosopher or historian. If all three groups of intellectuals would work together for the same end, so much the better.

Every scientist who reads the newspapers or listens to discussion knows that people try to influence political opinion in ways that would be universally condemned in his own sphere. Craftily misleading, confused, false, and sarcastic statements are made; facts are wilfully exaggerated or minimized; low motives are imputed without evidence; slogans are repeated; invective is used as a substitute for argument; hate is stirred up by the exhibition of hate; atrocities are condemned only when perpetrated by political opponents; the idea of truth as an end in itself is derided. These methods are by no means confined to the popular Press, but flourish also in political journals intended for intellectuals. The outlook of those who wish to affect public opinion by falsehood has been unequivocally stated by Mr David Low, the cartoonist, in the *New Statesman*. "Remember," he writes, "we are not here to pass honest judgments but to purvey matter the other side doesn't like." [68] Exactly. Politics is permeated by the outlook crystallized in these words. Scientists who care for truth outside as well as inside their own subject should use their influence to eradicate this outlook. It will be an uphill task, but not an impossible one. They should press for the introduction of their own methods of argument into politics.

The cinematograph is nowadays being used to affect public opinion without regard for truth. People who are little influenced by print, because they read little, can be greatly influenced by what they see upon the screen; and the opinion of many people who do read is much more easily influenced by the cinema than by other methods of propaganda. Films making propaganda for a foreign political regime profess to be truthful, but in fact not only give a false general impression, but actually represent incidents which are known never to have occurred. The menace of cinematographic propaganda seems not to have been realized. Scientists should not be content to let people absorb falsehoods by the prostitution of a marvellous instrument which owes its existence to science. The falsehoods contained in films which profess to represent the truth should be exposed directly the censorship permits it. (The cinematograph is also being prostituted in another way. Our ancestors made public executions illegal. The cinema now undoes their wise and civilizing legislation.)

Three arguments have been brought against me when I have suggested in public that scientists should try to improve the standards of political controversy. The first is not very important. Public speaking and writing, it is said, would become weak and spineless if the suggestions were put into practice. This is a misunderstanding. No objections can properly be raised to forcible wording, if the words contain no rhetorical device intended to affect opinion without the use of reason by the hearers or readers. Sentences may be decorated with metaphors, similes, and humour, as is sometimes done in scientific discussions, if the result is only to clarify and make vivid the meaning of what is being said and not to persuade others by subterfuge. Nothing can be urged against the straightforward condemnation of what is thought to be wrong, for if the speaker or writer expresses himself unequivocally, it is easy for him to be corrected when he makes a mistake (and we all make mistakes, in politics as well as in science).

The second criticism is serious. It is said that the common man neither uses nor wishes to use reason, and that therefore one should appeal only to his emotions. There is an inherent fallacy here, for the most reasonable man would be as lazy and unproductive as a pig if he were not inspired to action by emotion: both emotion and reason are necessary for reasonable action. There is no harm in seeking to arouse emotion in one's hearers or readers, if it is done only by a truthful presentation of the facts that have aroused the same emotion in oneself. Every successful science teacher does this in his lectures. The serious point is that the critics are willing that our democracy should be governed by the opinions of unreasonable people. Surely anyone who thinks that the common men of a democracy do not use reason should press urgently for the institution of anti-rhetoric classes in our elementary schools, instead of recommending that politicians should continue to use unsound arguments. It is questionable whether there is any greater single benefit that could be conferred on a democratic community' than the teaching of the principles of reasonable argument in every school.

A third criticism made against the suggestion that only reasonable arguments should be used in politics is that, whereas the general welfare of scientists is little affected by the theories they discuss, politicians are concerned with matters that deeply affect the day-to-day life of themselves and others. While the scientist naturally uses reason, it is argued, the politician rightly stirs up emotion by any available device. This argument is easily answered. The scientist uses reason in discussing his theories because he wishes to arrive at truth. In so far as the politician does not use reason, he does not wish to arrive at truth, but on the contrary intends to persuade people to accept and act upon untruth. He wishes, therefore, that the people should agree to legislation to which they would not have agreed if they had known the truth.

Famous research workers could use their great prestige to influence a wide public to demand higher standards in political controversy. Science masters at schools could exert a powerful influence for good if they were to suggest to their pupils that sound methods of argument are applicable in the discussion of matters affecting social life. Science students at universities could also be influential. They could attend the meetings of the political clubs of their universities, and methodically expose the illegitimate rhetorical devices used by the speakers. They could show what parts of the arguments were valid and of a kind that could reasonably affect opinion, and what parts were irrelevant. If that were done, the influence of those science students would spread far beyond their universities. Scientists of all kinds who want to make a special contribution to general welfare apart from the fulfilment of their primary duty would find here a congenial field of action.

It is, of course, far easier for the scientist than for the politician to use valid methods of argument, for the latter is subjected to severe temptations to put expediency before principle and to think of the desirability of the ends he seeks rather than of the justice of the means he uses to attain them. It may be true that on the average science attracts a higher type of man than politics, but if the scientist remembers his freedom from the difficulties that beset the politician, he will not lightly make the claim. Nevertheless, without any unwarrantable self-satisfaction, the scientist may legitimately point out that in science we have a means of arriving at truth which we should be happy to share with others. It seems probable that the historian of the future will find it hard to understand how twentieth-century science can have co-existed with the methods now employed in political discussion. He is likely to ask why the twentieth-century scientist was content to let legislation depend on methods of argument so palpably false, without even suggesting that the state of affairs could be improved.

Scientists have another political responsibility beyond those of urging the necessity for free speech and trying to reform the methods of political argument. Science is international, and there is a duty towards the scientists of other lands. In totalitarian states scientists see their subject subverted by political issues, are deprived of liberty of speech and publication, and are subject to wrongful dismissal, imprisonment, exile, or execution (see, *e.g.*, V. V. Tchernavin). [100] The scientist's feeling of world citizenship should forbid him to remain aloof. Something has indeed been done for some of the scientists who have suffered under Nazi tyranny, but little attention has been paid to sufferers under other totalitarian regimes. When making propaganda for totalitarianism, British scientists have sometimes callously overlooked the persecution of their fellow-workers. This field for humane action will remain open until scientists throughout the world have freedom of speech, publication, and inquiry, and are no longer subject to dismissal or punishment on account of race, class, or political beliefs.

5. *The Intellectual and the Common Man*

There is one duty that the scientist shares with every other kind of intellectual.

Human life has dignity mainly because men and women exist, and have existed, who are exceptional in virtue or intellect. They have arisen from every class. Honour should be accorded to these exceptional men and women, and to those who make an environment in home, school, or university in which genius and talent can develop and flourish.

Many people to-day think of progress as meaning one thing only, the improvement of the material conditions of the poorer members of the community. This is a limited outlook, for true progress means something more than material advancement. It means movement towards greater things in science, philosophy, art, music, literature, and all other branches of intellectual activity. That kind of progress is perfectly compatible with the securing of a square deal and a happy life for the common man, and conflict can only arise when the needs of the common man are made paramount.

The common man is very willing to throw his daily penny into the coffers of the newspaper proprietor who will tell him sufficiently regularly what an uncommonly fine fellow he is, and how paramount are his interests. He does not recognize the logical absurdity of the praise. He does not understand that, if he were in any way uncommon in virtue or talent, he would not belong to the group of people on whom the praise is bestowed. The false attachment of special virtue and talent to people who by definition do not possess them has become so widespread as to constitute a serious threat to civilization. People are beginning to cease honouring great men, and to honour instead the masses of humanity or the publicists who reiterate their praise.

It can never have entered the heads of J. S. Mill and the other great apostles of liberty that the very people who owe their liberty to them would seek to destroy the gift. Nevertheless, it seems that the common man has not an urgent desire for liberty of action, and is prepared to use the vote granted to him by liberal-minded people to destroy not only his own liberty, but that of uncommon people as well. It is impossible to imagine that the common man understands the conditions under which great work in science, philosophy, or music can be done: he is prepared and actively encouraged to think that the only thing that matters is his own material welfare. Further, he is apt to think that he has only to hand over the control of the affairs of the nation to a central planner and his economic welfare will be assured. The central planner has told him so.

That way lies the eclipse of intellectual life. Every thoughtful person has a duty to strive to prevent it. Not a few intellectuals, however, are doing exactly the opposite. So acutely conscious are they that common men have their rights and legitimate aspirations, that they fail to recognize the imperative necessity to prevent the satisfaction of these rights and aspirations from interfering with those of uncommon men. If throughout the centuries we had only had common men, we should still be living like savages. Progress will be slow if the wants of common men are made paramount. Those intellectuals who ally themselves with the herd to act against the interests of uncommon men are guilty of intellectual treason.

6. *Duties towards Animals*

These, then, are kinds of services that scientists can render to society without impairment of their integrity: they can protect the heritage of science against those who would destroy it; they can extend the appreciation of science among a wider proportion of the community; they can facilitate the entry of suitable people of all classes into professional science; they can use their influence in favour of freedom of speech and urge better methods of political argument; they can serve the interests of their fellow-workers in totalitarian states; and they can use their influence against the idea that the interests of the common man are paramount. These are services to their fellow-men, but their potentiality for usefulness does not end there. As J. S. Mill [72] pointed out, the standard of morality should not be confined in its application to human beings, but should be secured, "so far as the nature of things admits, to the whole sentient creation."

The word *sentient* in this context presumably means conscious. Here a perplexity assails us. Everyone knows, more certainly than he knows anything else whatever, that he is conscious, but he can only infer that anyone else is. Similarly he can only infer, and never be certain, that the higher animals have consciousness. In so far, however, as he treats other human beings with consideration because he believes them to be conscious, he should also act kindly towards those animals about which he draws the same inference. No one who has studied the nervous system and reactions of a sea-anemone would be likely to think it possible to be "cruel" to such a lowly animal, lacking as it does any central nervous system or brain. One might as well think of cruelty to one's own intestines, which may be cut freely by the surgeon without the use of anæsthetics and without the feeling of pain, though they have a diffuse nervous system of their own which is capable of ordering some quite complex responses to stimuli. When, however, one studies the brains of some of the higher animals, particularly the higher Vertebrates, and notes their reactions to complex situations, one can only say that it seems very probable that they are conscious and capable of enjoyment and suffering. We are here on grounds not of certainty, but of that degree of probability which determines moral judgments.

The biologist knows far better than others what a remarkable resemblance there is between the nervous system of a higher Vertebrate and that of man, and how certain centres of the brains of both are regarded as pain centres. It therefore devolves upon him, more than on the man in the street, to encourage kindness to animals.

A distinguished physiologist (I think it was Professor J. B. S. Haldane, F.R.S.) has somewhere remarked that he has never known a good physiologist who was fond of shooting as a sport. It is, I believe, true that if people knew more about the nervous systems of animals, they would be less inclined to take up shooting for sport, on account of the suffering caused to wounded mammals and birds. It is no good pressing for the abolition of shooting: that would only cause devotees of the sport and the business interests concerned to bring powerful influence to bear in the contrary direction. Education of the young is the only hopeful way of tackling the subject. The present situation is not satisfactory. Men are sent to prison for causing suffering to tame animals, but they are perfectly free to cause much more suffering to wild animals. The trapping of animals for their furs probably causes more suffering than shooting, and some of the methods used are made particularly painful by the necessity not to damage the skin. Here again it is useless to try direct opposition against a powerful industry and the vanity and thoughtlessness of women, many of whom would probably give up wearing furs if they had once seen a living fur-bearer held in a trap. Biological education is again the remedy, reinforced by a law insisting that every skin offered for sale should bear a label stating whether the animal was trapped or bred on a fur-farm.

The killing of animals for food has become much more humane in recent times, but the methods of castration and ovariotomy require investigation, especially the latter.

Although biologists bear a responsibility to educate the public in the matter of the treatment of animals generally, clearly they have a very special responsibility for the animals upon which their own work is performed. Although the housing arrangements for animals are excellent in some laboratories, yet this is by no means always true. The Universities Federation for Animal Welfare is to be congratulated on the steps that it is taking on behalf of laboratory animals. The Federation is compiling information on methods of housing, feeding, anæsthetizing, and killing. [105] The vexed question of vivisection is one on which the scientist cannot remain indifferent. The public probably does not realize that a large number of so-called vivisection experiments involve no more than changing the diet or making subcutaneous injections, and that the animal on which a major operation has been performed under anæsthesia is often killed without being allowed to regain consciousness. It is a fact that physiology would be struck an almost mortal blow if vivisection were made illegal, and the influence on medicine (human and veterinary) would be disastrous. Nevertheless, scientists have a duty to exercise the utmost thoughtfulness on this subject. I believe that painful experiments are sometimes performed, of a kind that is extremely unlikely to give valuable new knowledge. The scientist should take a lot of trouble in planning his experiments so as to reduce pain to a minimum. Meanwhile it is a healthy sign that societies exist to protect animals against vivisection. It is probable, however, that some members of such societies have a rather distorted view of the subject. I have never come across a case in which a scientist practised vivisection because he found pleasure in cruelty.

A point worth mentioning in connexion with this subject is that invertebrate animals are not legally protected. It might be desirable to modify the law. Some of the invertebrates, such as the octopus and its allies, have astonishingly highly developed sensory and nervous systems. The redrafting of the law, however, would be very difficult.

7. *Conclusion*

The purpose of this chapter has been to suggest various ways in which the scientist may use his special knowledge, talents, and outlook to make the world a better place, apart from the fulfilment of his primary duty in research or teaching. None of the suggestions made involves any sacrifice of the spirit of science. During the last twelve years much has been said about the social responsibilities of scientists, but nearly all of it has been quite different from what has been said in this chapter. Scientists have been urged to regard their subject as existing solely for service to man's material wants, to press for the central planning of scientific research, and to ally themselves with political groups which advocate the central planning of society in general. Scientists should not accept this advice, for three reasons:

First, science does not exist solely to serve man's material wants.

Secondly, any thoroughgoing scheme for the central planning of research would gravely damage science.

Thirdly, totalitarianism is precisely the form of government that is least in accord with scientific principles; for scientists accept the authority of no one and recognize the necessity for liberty.

These three clauses are the answers to the contrary arguments stated in Chapter I. The purpose of this book has been to give the reasons why these answers are valid, and to suggest social responsibilities that scientists can undertake without sacrifice of the ideals they ought to serve. By undertaking these responsibilities in addition to their primary duties, scientists can show the reality of their belief in the liberty, fraternity, and inequality of man.

List of References

1. ABRAHAM, E. P., E. CHAIN *et al.* 1941. "Further Observations on Penicillin," *Lancet*, p. 177.

2. ADAMS, M. (edited by). 1933. *Science in the Changing World.* George Allen & Unwin, London.

3. ÅNON. 1787. *Mémoire historique sur la vie et les écrits de Monsieur Abraham Trembley.* Fauche, Neuchâtel.

4. ANSHEN, R. N. (edited by). 1942. *Freedom, its Meaning.* George Allen & Unwin, London.

5. BAAMAN, E., and E. RIEDEL. 1934. "Ueber das Vorkommen zweier durch das pH-Wirkungsoptimum unterscheidbaren Phosphoesterasen in tierischen Organ," *Zeit. physiol. Chem.*, 229, p. 125.

6. BAKER, J. R. 1942. *The Scientific Life.* George Allen & Unwin, London.

7. BAKER, J. R. 1942. Letter in *Endeavour*, 1, p. 90.

8. BERNAL, J. D. 1939. *The Social Function of Science.* Routledge, London.

9. BERNAL, J. D. 1940. "Science in the U.S.S.R.," *New Statesman and Nation*, 24th February, p. 243.

10. BERNAL, J. D. 1942. Chapter on "Present-day Science and Technology in the Soviet Union," in Needham and Davies.[76]

11. BOHR, N. 1940. Foreword to Moller and Rasmussen.[74]

11a. BRAEM, F. 1890. "Untersuchungen über die Bryozoen des süssen Wassers," *Bibl. zool.*, 2 (6), p. 1.

12. BRIDGMAN, P. W. 1943. "Science, and its Changing Social Environment," *Science*, 97, p. 147.

13. British Association. 1942. *Science and World Order.* Published by the Association, London.

14. BUKHARIN, N. I., *et al.* 1935. *Marxism and Modern Thought.* Translation by R. Fox. Routledge, London.

15. CANTERBURY, ARCHBISHOP OF. 1943. Quoted in *Sunday Express*, 3rd October.

16. CHAIN, E., H. W. FLOREY *et al.* 1940. "Penicillin as a Chemotherapeutic Agent," *Lancet*, p. 226.

17. CONKLIN, E. G. 1993. "Predecessors of Schleiden and Schwann," *Amer. Nat.*, 73, p. 538.

18. CROWTHER, J. G. 1936. *Soviet Science.* Kegan Paul, Trench, Trübner & Co., London.

19. CROWTHER, J. G. 1941. *The Social Relations of Science* Macmillan, London.

20. CROWTHER, J. G. 1942. "Science in Soviet Russia," *Nature*, 150, p. 647.

21. CROWTHER, J. G. 1942. Review of *The Scientific Life* (by J. R. Baker) in *New Statesman and Nation*, 15th August.

22. DARWIN, C. 1872. *The Origin of Species by means of Natural Selection, or the preservation of favoured races in the struggle for life.* 6th edition. Murray, London.

23. DARWIN, F. 1903. *More Letters of Charles Darwin: a record of his work in a series of hitherto unpublished letters.* Vol. 1. Murray, London.

24. DAVIS, D. R. 1934. "The Phosphatase Activity of Spleen Extracts," *Biochem. Journ.*, 28, p. 529.

25. Editorial. 1944. "Control of Cancer," *Lancet*, 1st January, p. 21.

26. EINSTEIN, A. 1942. Chapter on "Freedom and Science" in Anshen.[4]

27. ELLIS, J. 1755. *An Essay towards a Natural History of the Corallines, and other Marine Productions of the like kind, commonly found on the coasts of Great Britain and Ireland.* Published by the author, London.

28. ELLIS, J., quoted by Johnston.[55]

29. 'ESPINASSE, P. G. 1941. "Genetics in the U.S.S.R.," *Nature*, 148, p. 739.

30. FARADAY, M., quoted by Jones.[56]

31. FARADAY, M., quoted by Gregory.[35]

32. FLEMING, A. 1929. "On the Antibacterial Action of Cultures of a Penicillium, with special reference to their use in the isolation of B. influenzæ," *Brit. Journ. Exp. Path.*, 10, p. 226.

33. FRÉDÉRICQ, M. L. 1890. "Sketch of Theodor Schwann," *Pop. Sci. Mon.*, 37, p. 257.

34. GARRISON, F. H. 1929. *An Introduction to the History of Medicine.* Saunders, London.

35. GREGORY, R. A. 1916. *Discovery, or the spirit and service of Science.* Macmillan, London.

36. GREGORY, R. (A.). 1941. "The Commonwealth of Science," *Nature*, 148, p. 393.

36a. GREGORY, R. (A.). 1941. *Science in Chains.* Macmillan, London.

37. GRIFFITHS, J. 1889. "Observations on the Function of the Prostate Gland in Man and the Lower Animals," *Journ. Anat. Physiol.*, 24, p. 27.

38. GRIJNS, G. 1901. See Mathews.[70]

39. GROSSER, P., and J. HUSLER. 1912. "Über das Vorkommen einer Glycerophosphotase in tierischen Organen," *Biochem. Zeit.*, 39, p. 1.

40. GRUBER, W. 1847. "Untersuchung einiger Organe eines Castraten," *Arch. Anat., Physiol. u. wiss. Med.*, 6, p. 463.

41. GUTMAN, A. B. and E. B. 1938. "An 'Acid' Phosphatase occurring in the serum of patients with Metastasizing Carcinoma of the Prostate Gland," *Journ. Clin. Invest.*, 1, p. 473.

42. HALDANE, J. B. S. 1933. Chapter on "The Biologist and Society" in Adams.[2]

43. HARDY, G. H. 1941. *A Mathematician's Apology.* Cambridge University Press.

44. HARROW, B., and C. P. SHERWIN. 1934. *The Chemistry of the Hormones.* Baillière, Tindall & Cox, London.

45. HEITZ, E., and H. BAUER. 1933. "Beweise für die Chromosomennatur der Kernschleifen in der Knäuelkernen von Bibio hortulanus," *Zeit. Zellforsch. u. mikr. Anat.*, 17, p. 67.

46. HOGBEN, L. 1942. "Biological Instruction and Training for Citizenship," *School Sci. Rev.*, 91, p. 263.

47. HOLDER, C. F. 1893. *Louis Agassiz, his Life and Work.* Putman, London.

48. HOPKINS, F. G. 1933. "Some Chemical Aspects of Life," *Rep. Brit. Ass. Adv. Sci.*, 103, p. 1.

49. HUGGINS, C. 1943. "Endocrine Control of Prostatic Cancer," *Science*, 97, p. 541.

50. HUMBOLDT, A. VON. 1849. *Cosmos: a sketch of a physical description of the Universe.* Translation by E. C. Otté. Bohm, London.

51. HUNTER, J., quoted by Griffiths.[37]

52. HUXLEY, J. S., and G. R. DE BEER. 1934. *The Elements of Experimental Embryology.* Cambridge University Press.

53. HUXLEY, T. H. 1894. *Method and Results.* Macmillan, London.

54. INFELD, L. 1941. *Quest: the Evolution of a Scientist.* Gollancz, London.

55. JOHNSTON, G. 1847. *A History of British Zoophytes.* 2 vols. Van Voorst, London.

56. JONES, B. 1870. *The Life and Letters of Faraday.* 2 vols. Longmans, Green, London.

57. KAPITSA, P. L. 1942. "Science and War," *Science*, 95, p. 396.

58. KARPOVA, L. 1925. "Beobachtungen über den Apparat Golgi (Nebenkern) in den Samenzellen von Helix pomatia," *Zeit. Zellforsch. u. mikr. Anat.*, 2, p. 495.

59. KITCHIN, G. 1935. *Prisoner of the OGPU.* Longmans, Green, London.

60. KOLBANOVSKY, V., *et al.* 1939. "Conference on Genetics and Selection: controversial questions in Genetics and Selection" (in Russian)," *Pod Znamenem Marxisma*, 11, p. 86.

61. KOPSCH, F. 1902. "Die Darstellung des Binnennetzes von Golgi in spinalen Ganglienzellen und anderen Körperzellen mittels Osmiumsäure," *Sitz.-Ber. Akad. wiss. Berl.*, 39.

62. KOSSEL, W. 1916. "Über Molekülbildung als Frage des Atomsbaus," *Ann. der Physik*, 49, p. 229.

63. KUTCHER, W., and H. WOLBERGS. 1935. "Prostataphosphatase," *Zeit. physiol Chem.*, 236, p. 237.

64. LACASSAGNE, A. 1933. "Métaplasie épidermoïde de la prostata provoquée, chez la souris, par des injections répétées de fortes dóses de folliculine," *Compt. rend. Soc. biol.*, 113, p. 590.

65. LAGRANGE, J. L., quoted by Turner. [104]

66. LANGMUIR, I. 1919. "The Arrangement of Electrons in Atoms and Molecules," *Journ. Amer. Chem. Soc.*, 41, p. 868.

66a. LEWIS, G. N. 1916. The Atom and the Molecule," *Journ. Amer. Chem. Soc.*, 38, p. 762.

67. LIPPMANN, W. 1938. *The Good Society.* George Allen & Unwin, London.

68. LOW, D. 1943. Letter in *New Statesman and Nation*, 31st July, p. 74.

69. LYONS, E. 1938. *Assignment in Utopia.* Harrap, London.

70. MATHEWS, A. P. 1939. *Physiological Chemistry: a text-book for students.* Ballière, Tindall & Cox, London.

71. McCLUNG, C. E. 1902. "The accessory chromosome—sex determinant?", *Biol. Bull.*, 3, p. 43.

72. MILL, J. S. 1863. *Utilitarianism.* Reprinted in *Utilitarianism, Liberty, and Representative Government* in Everyman's Library. Dent, London.

73. MILLER, H. 1841. *The Old Red Sandstone.* Reprinted in Everyman's Library. Dent, London.

74. MØLLER, C., and E. RASMUSSEN. 1940. *The World and the Atom.* George Allen & Unwin.

75. NASSONOV, D. N. 1923. "Das Golgische Binnennetz und seine Beziehungen zu der Sekretion. Untersuchungen über einige Amphibiendrüsen," *Arch. f. mikr. Anat.*, 97, p. 136.

76. NEEDHAM, J., and J. S. DAVIES (edited by). 1942. *Science in Soviet Russia.* Watts, London.

77. PAINTER, T. S. 1934. "Salivary Chromosomes and the Attack on the Gene," *Journ. Hered.*, 25, p. 465.

78. PAVLOVSKY, E. N. 1917. *Materials on the Comparative Anatomy and Development of the Scorpions.* Petrograd. Quoted by Pavlovsky and Zarin.[80]

79. PAVLOVSKY, E. N. 1926. "Studies on the Organization and Development of Scorpions: 5. The Lungs," *Quart. Journ. Micr. Sci.*, 70, p. 135.

80. PAVLOVSKY, E. N., and E. J. ZARIN. 1926. "On the Structure and Ferments of the Digestive Organs of Scorpions," *Quart. Journ. Micr. Sci.*, 70, p. 221.

81. POINCARÉ, H. 1934. *Science et méthode.* Flammarion, Paris.

82. POLANYI, M. 1941. Lecture at Leeds University on "The Social Message of Science."

83. POLANYI, M. 1941. "The Growth of Thought in Society," *Economica*, 8, p. 428.

84. POLANYI, M. 1943. "The Autonomy of Science," *Mem. Proc. Manch. Lit. Phil. Soc.*, 85, 2, p. 19.

85. PREZENT, I. I., quoted by Doubinin; see Kolbanovsky *et al.* [60]

86. RAMON Y CAJAL, S. 1912. "Fórmula de fijación para la demonstración facil del aparato reticolar de Golgi," *Trab. Lab. Invest. Biol.*, 10, p. 209.

87. RAUSCHNING, H. 1939. *Hitler speaks: a series of Political Conversations with Adolph Hitler on his real aims.* Butterworth, London.

88. RÉAUMUR, M. DE. 1742. *Mémoires pour servir à l'histoire des Insectes.* Vol. 6. Imprimerie Royale, Paris.

89. ROUDABUSH, R. L. 1933. "Phenomenon of Regeneration in everted Hydra," *Biol. Bull.*, 64, p. 253.

90. ROWSE, A. L. 1942. *A Cornish Childhood.* Cape, London.

91. RUSSELL, B. 1941. *Let the People think: a Selection of Essays.* Watts, London.

91a. SAX, K. 1944. "Soviet Biology," *Science*, 99, p. 298.

92. SHOENBERG, D. 1942. Chapter on "Physical Research in the Soviet Union" in Needham and Davies. [76]

92a. STAMP, J. 1936. "The Impact of Science upon Society," *Brit. Ass. Adv. Sci. Rep., Blackpool*, p. 1.

93. STEPHENSON, J. 1931. "Robert Brown's discovery of the Nucleus in relation to the history of the Cell Theory," *Proc. Linn. Soc.*, 144, p. 45.

94. STEINACH, E., and H. KUN, quoted by Lacassagne. [64]

95. SULLIVAN, J. W. N. 1938. *Isaac Newton, 1642-1727.* Macmillan, London.

96. SUTTON, W. S. 1902. "On the Morphology of the Chromosome Group in Brachystola magna," *Biol. Bull.*, 4, p. 24.

97. SWANN, quoted in *Nature*, 150, p. 383.

98. SZENT-GYÖRGYI, A. 1943. "Science needs Freedom," *World Digest*, 55, p. 50.

99. TCHERNAVIN, T. 1933. *Escape from the Soviets.* Hamish Hamilton, London.

100. TCHERNAVIN, V. V. 1935. *I speak for the Silent Prisoners of the Soviets*. Hamish Hamilton, London.

101. THOMPSON, D'A. W. 1917. *On Growth and Form*. Cambridge University Press.

102. TREMBLEY, A. 1744. *Mémoires, pour servir à l'histoire d'un genre de polypes d'eau douce, à bras en forme de cornes*. J. and H. Verbeek, Leide.

103. TRUEMAN, A. E. 1943. *Science and the Future*. The British Way Series. Craig Wilson, Glasgow.

104. TURNER, D. M. 1927. *Makers of Science: Electricity and Magnetism*. Oxford University Press.

105. Universities Federation for Animal Welfare. 1943. "Welfare of Laboratory Animals," *Nature*, 151, p. 723.

106. VALLERY-RADOT, R. 1900. *La vie de Pasteur*. Hachette, Paris.

107. WALTON, A. 1942. Chapter on "Soviet Agricultural Science" in Needham and Davies. [76]

108. WESTAWAY, F. W. 1942. *Science in the Dock: Guilty or Not Guilty?* Blackie, London.

109. WHEWELL, W. 1840. *The Philosophy of the Inductive Sciences*. Parker, London.

110. WILSON, E. B. See references to his own work given in his book, *The Cell in Development and Heredity*, 3rd edition (1925). Macmillan, New York.

Index

HISTORY, PHILOSOPHY AND
SOCIOLOGY OF SCIENCE

Classics, Staples and Precursors

An Arno Press Collection

Aliotta, [Antonio]. **The Idealistic Reaction Against Science.** 1914

Arago, [Dominique François Jean]. **Historical Eloge of James Watt.** 1839

Bavink, Bernhard. **The Natural Sciences.** 1932

Benjamin, Park. **A History of Electricity.** 1898

Bennett, Jesse Lee. **The Diffusion of Science.** 1942

[Bronfenbrenner], Ornstein, Martha. **The Role of Scientific Societies in the Seventeenth Century.** 1928

Bush, Vannevar. **Endless Horizons.** 1946

Campanella, Thomas. **The Defense of Galileo.** 1937

Carmichael, R. D. **The Logic of Discovery.** 1930

Caullery, Maurice. **French Science and its Principal Discoveries Since the Seventeenth Century.** [1934]

Caullery, Maurice. **Universities and Scientific Life in the United States.** 1922

Debates on the Decline of Science. 1975

de Beer, G. R. **Sir Hans Sloane and the British Museum.** 1953

Dissertations on the Progress of Knowledge. [1824]. 2 vols. in one

Euler, [Leonard]. **Letters of Euler.** 1833. 2 vols. in one

Flint, Robert. **Philosophy as Scientia Scientiarum and a History of Classifications of the Sciences.** 1904

Forke, Alfred. **The World-Conception of the Chinese.** 1925

Frank, Philipp. **Modern Science and its Philosophy.** 1949

The Freedom of Science. 1975

George, William H. **The Scientist in Action.** 1936

Goodfield, G. J. **The Growth of Scientific Physiology.** 1960

Graves, Robert Perceval. **Life of Sir William Rowan Hamilton.** 3 vols. 1882

Haldane, J. B. S. **Science and Everyday Life.** 1940

Hall, Daniel, et al. **The Frustration of Science.** 1935

Halley, Edmond. **Correspondence and Papers of Edmond Halley.** 1932

Jones, Bence. **The Royal Institution.** 1871

Kaplan, Norman. **Science and Society.** 1965

Levy, H. **The Universe of Science.** 1933

Marchant, James. **Alfred Russel Wallace.** 1916

McKie, Douglas and Niels H. de V. Heathcote. **The Discovery of Specific and Latent Heats.** 1935

Montagu, M. F. Ashley. **Studies and Essays in the History of Science and Learning.** [1944]

Morgan, John. **A Discourse Upon the Institution of Medical Schools in America.** 1765

Mottelay, Paul Fleury. **Bibliographical History of Electricity and Magnetism Chronologically Arranged.** 1922

Muir, M. M. Pattison. **A History of Chemical Theories and Laws.** 1907

National Council of American-Soviet Friendship. **Science in Soviet Russia: Papers Presented at Congress of American-Soviet Friendship.** 1944

Needham, Joseph. **A History of Embryology.** 1959

Needham, Joseph and Walter Pagel. **Background to Modern Science.** 1940

Osborn, Henry Fairfield. **From the Greeks to Darwin.** 1929

Partington, J[ames] R[iddick]. **Origins and Development of Applied Chemistry.** 1935

Polanyi, M[ichael]. **The Contempt of Freedom.** 1940

Priestley, Joseph. **Disquisitions Relating to Matter and Spirit.** 1777

Ray, John. **The Correspondence of John Ray.** 1848

Richet, Charles. **The Natural History of a Savant.** 1927

Schuster, Arthur. **The Progress of Physics During 33 Years (1875-1908).** 1911

Science, Internationalism and War. 1975

Selye, Hans. **From Dream to Discovery: On Being a Scientist.** 1964

Singer, Charles. **Studies in the History and Method of Science.** 1917/1921. 2 vols. in one

Smith, Edward. **The Life of Sir Joseph Banks.** 1911

Snow, A. J. **Matter and Gravity in Newton's Physical Philosophy.** 1926

Somerville, Mary. **On the Connexion of the Physical Sciences.** 1846

Thomson, J. J. **Recollections and Reflections.** 1936

Thomson, Thomas. **The History of Chemistry.** 1830/31

Underwood, E. Ashworth. **Science, Medicine and History.** 2 vols. 1953

Visher, Stephen Sargent. **Scientists Starred 1903-1943 in American Men of Science.** 1947

Von Humboldt, Alexander. **Views of Nature: Or Contemplations on the Sublime Phenomena of Creation.** 1850

Von Meyer, Ernst. **A History of Chemistry from Earliest Times to the Present Day.** 1891

Walker, Helen M. **Studies in the History of Statistical Method.** 1929

Watson, David Lindsay. **Scientists Are Human.** 1938

Weld, Charles Richard. **A History of the Royal Society.** 1848. 2 vols. in one

Wilson, George. **The Life of the Honorable Henry Cavendish.** 1851